Situated Aesthetics

Art Beyond the Skin

SITUATED AESTHETICS

ART BEYOND THE SKIN

Edited By

Riccardo Manzotti

imprint-academic.com

This collection copyright © Riccardo Manzotti, 2011
Individual contributions © their authors, 2011

Cover illustration: La Scala Virtuale, © Federica Marangoni
Cover design: Concept Graphics, www.conceptstudio.co.uk

The moral rights of the author have been asserted
No part of this publication may be reproduced in any form
without permission, except for the quotation of brief passages
in criticism and discussion.

Published in the UK by Imprint Academic
PO Box 200, Exeter EX5 5YX, UK

Published in the USA by Imprint Academic
Philosophy Documentation Center
PO Box 7147, Charlottesville, VA 22906-7147, USA

ISBN 978 184540 238 9

A CIP catalogue record for this book is available from the
British Library and US Library of Congress

Contents

Acknowledgments vii
Notes on Authors viii

1. Riccardo Manzotti (Editor)
   Preface 1

Part One
From Externalism to Aesthetics

2. Riccardo Manzotti
   Varieties of Externalism and Aesthetics 15
3. Erik Myin and Johan Veldeman
   Externalism, Mind, and Art 37
4. Joel Krueger
   Enacting Musical Content 63
5. Liliana Albertazzi
   Extended Space in Perception and Art 87

Part Two
Externalist Approaches to Aesthetics

6. Robert Pepperell
   Art and Extensionism 107
7. Lambros Malafouris
   The Aesthetics of Material Engagement 123
8. Sabine Marienberg
   Language, Rhythm, Grain of Voice 141
9. Paola Carbone
   An Externalist Approach to Literature: From Novel to Cave Writing . 155

Part Three
Art Beyond the Skin

10. Teed Rockwell
    Musical Experience and the Extended Self 173
11. Sylvain Le Groux and Paul F.M.J. Verschure
    Music is Everywhere: A Situated Approach to Music Composition . . 189
12. Stéphane Dumas
    Creation as Secretion: An Externalist Model in Aesthetics 211
13. Giuliano Galletta
    The Self is Around Me 233

Index 241

Acknowledgments

I would like to extend my sincere gratitude to the persons and organizations that made possible both this volume and the conference that inspired it. In particular, I express my gratitude to the European Science Foundation (ESF) whose support permitted the publication of this volume.

The publication of this volume is counterfactually related to the occurrence of a curious situation which merits, I think, a word or two. The original commitment of many of the participants revolved around the possibility of providing a more scientific approach to aesthetics. By and large, the general feeling was that time was ripe to corroborate established and respected aesthetic theories with state-of-the-art models of the mind. The natural candidate was offered by neuroaesthetics pioneered by Semir Zeki's seminal and brilliant work. As a result neuroaesthetics prominently showed in the conference title. Nevertheless, during some of the most lively, intellectually rewarding, and interesting discussions, an alternative view gained momentum. Since a relevant number among the participants found such a view fascinating and promising, it was considered an opportunity to present this approach in a more systematic way. Of course, I am referring to externalism and its possible consequences for aesthetics. It must be stressed that it was neither possible nor desirable to put together several authors on a daring target like that and to expect them to share a very detailed picture of the forthcoming view. However, as it happens, in the end, the editor and the authors managed to get together by persuasion or charm to achieve a final work. I would like to remember that in science and in philosophy as well, there is what the Greek conductor and pianist Dimitri Mitropoulos used to call 'the sportive element' — namely a factor of curiosity, adventure, and experiment that makes a book worthy of being read, discussed, and perhaps agreed upon. It's in this spirit of adventure that this book is now published.

As the editor I must thank all the contributors of the book who bravely agreed to participate in a rather perilous intellectual enterprise whose precise goals and ends are not easy to foresee at the present time.

From the Institute of Communication and Behavior at IULM University in Milan, I thank the chair Professor Paolo Moderato for his continuous support, advice and encouragement. Finally, and not least, I want to express my gratitude to Professor Giovanni Puglisi, dean of IULM University, for creating and maintaining an intellectually open and rewarding environment where new ideas can freely cross the traditional watertight boundaries of academic trenches.

Riccardo Manzotti
January 2011

Notes on the Authors

Liliana Albertazzi is associate professor at Trento University and member of CIMeC (Centre for Mind & Brain, Rovereto, Italy). Her most recent works concern the nature of the perceptual and pictorial spaces of vision, the perceptual base of linguistic universals, the cognitive structure of metaphorical thinking, colour perception and colour categorization, information in perception, and the origin of meaning from grouping and shape configuration. Recently she edited the volume *Visual Thought: The Depictive Space of Perception* (John Benjamins, Amsterdam) and *Perception Beyond Inference: The Information Content of Perceptual Processes* (MIT Press, Cambridge, MA).
liliana.albertazzi@unitn.it

Paola Carbone is Associate Professor of English Literature at IULM University, Milan. Her fields of research include: narrative theory; contemporary British culture and novel; the relationship between literature and new communication technologies. She has published several works on postmodern literature and digital art. In 2008 she published a groundbreaking analysis of Laurence Sterne's work. Since 2000 she has coordinated the Tristram Shandy Web Project (http://www.tristramshandyweb.it/). Her present research focuses on narration and the emergence of mental images.
Paola.carbone@iulm.it

Stéphane Dumas is an art theoretician and a visual artist. As a theoretician, he has written papers about the problem of embodiment in art, the status of image as a 'creative skin', and the aesthetics of liminality. A book entitled *Creative Skins* is in preparation. As an artist, he works on the fragmented human figure and the skin. His work has been shown in numerous places, and is included in museum collections. He teaches at ESAA Duperré, Paris, and is part of a research laboratory in aesthetics at Sorbonne University.
stedumas@free.fr

Acknowledgments

Giuliano Galletta (www.giulianogalletta.it) is an Italian artist, writer and journalist. He was born in Sanremo in 1955 and currently lives in Genoa where he is a staff writer at the Italian newspaper 'Il Secolo XIX'. He has presented his work in several galleries and art museums both in Italy and abroad. Among his most recent exhibitions are 'The Chaos Museum', at the Villa Croce Museum of Contemporary Art in Genoa. He is the author of many books on the human condition. The most recent is *The World is Not a Peach* (Socialmente, 2010).
galletta@ilsecoloxix.it

Joel Krueger is a postdoctoral research fellow at the Center for Subjectivity Research, University of Copenhagen. His current research focuses on the embodied and enactive roots of social cognition and music perception. He has published articles on various issues in phenomenology and philosophy of mind, Asian and comparative philosophy, pragmatism, and philosophy of music.
joelk@hum.ku.dk

Sylvain Le Groux is a researcher at the laboratory for Synthetic, Perceptive, Emotive and Cognitive Systems of Pompeu Fabra University in Barcelona. He is also an active musician and interaction designer. He builds and evaluates synthetic interactive systems to address questions at the intersection of perception, cognition, emotion, therapy, and performance.
sylvain.legroux@upf.edu   http://www.dtic.upf.edu/~slegroux

Lambros Malafouris, PhD (Cambridge), is a Fellow in Creativity at Keble College, University of Oxford, and a former Balzan Research Fellow in Cognitive Archaeology at the McDonald Institute, University of Cambridge. His research interests lie broadly in the archaeology of mind and the philosophy of material culture. His recent publications include *The Cognitive Life of Things: Recasting the Boundaries of the Mind* (with Colin Renfrew), *Material Agency: Towards a Non-Anthropocentric Approach* (with Carl Knappett) and *The Sapient Mind: Archaeology Meets Neuroscience* (with Colin Renfrew and Chris Frith).
lambros.malafouris@keble.ox.ac.uk

Riccardo Manzotti is currently Professor of Psychology at IULM University in Milan. His main interests are the nature of consciousness and the design and implementation of models of conscious agents. He is a lecturer in Psychology of Art, and Neuroscience of Perception. He has a degree in Philosophy and another in Electronic Engineering. He has a PhD in Robotics focusing on Artificial Intelligence and models of Artificial Consciousness and Goal-Driven Artificial Agents.

He has published several papers on consciousness, externalism, and

ontological issues as to the nature of phenomenal experience in a physical world. He edited a book on the topic of artificial consciousness and, more recently, a volume on externalism and aesthetics.
Riccardo.manzotti@iulm.it

Sabine Marienberg studied Romance Studies and Philosophy in Munich, Perugia and Berlin and holds a PhD in Philosophy and Humanities from the Freie Universität Berlin. She is currently a lecturer in Philosophy at the Humboldt Universität Berlin and member of the interdisciplinary research group Funktionen des Bewusstseins at the Berlin-Brandenburgische Akademie der Wissenschaften. Her research interests lie in the areas of Philosophy of Language, Philosophy of Mind, Philosophical Anthropology and Aesthetics.
marienberg@gmx.net

Erik Myin teaches and conducts research at the Department of Philosophy of the University of Antwerp, where he is director of the Centre for Philosophical Psychology. His area of interest is the philosophy of cognitive science, often with a focus on perception. He has published on issues ranging from spectrum inversion to sensory substitution, sometimes in collaboration with working scientists, in places like Synthese, Journal of Consciousness Studies, Cognitive Science or The Cambridge Handbook of Situated Cognition. With Dan Hutto, he is currently working on a book manuscript titled *Radicalizing Enactivism* (under contract with MIT Press).
Erik.Myin@ua.ac.be

Robert Pepperell is an artist and writer. Trained at the Slade School of Art, UCL, London, he worked as a multimedia and installation artist through the 1990s with exhibitions at The Barbican, the ICA, the Millennium Dome, Glasgow Gallery of Modern Art, Ars Electronica, and others. He has written several books, including *The Posthuman Condition: Consciousness Beyond the Brain* (1995/2003), exhibits his paintings and drawings regulary, and is currently Professor of Fine Art at Cardiff School of Art in the UK.
rpepperell@uwic.ac.uk

Teed Rockwell is in the Philosophy department at Sonoma State University. He is the author of the MIT press book *Neither Brain nor Ghost*, which defends a radically externalist view of mind inspired by the American pragmatists. He is also the only person who performs Indian Ragas on a new instrument called the Touchstyle Veena. His other philosophical writings can be found at www.cognitivequestions.org and his music can be heard at www.myspace.com/teedrockwell
teedrockwell@gmail.com

Johan Veldeman is a researcher at the Department of Philosophy and the Centre for Philosophical Psychology at the University of Antwerp. His research interests are in consciousness, pictorial representation, aesthetic experience, and philosophy of art. His publications include 'Reconsidering Pictorial Representation by Reconsidering Visual Experience' (*Leonardo*, 2008) and 'Varieties of Phenomenal Externalism' (*Teorema*, 2009). *johan.veldeman@ua.ac.be*

Paul Verschure is a research professor with the Catalan Institute of Advanced Studies (ICREA) and the Universitat Pompeu Fabra. Paul uses synthetic and experimental methods to find a unified theory of mind and brain and applies the outcomes to novel real-world technologies and quality of life enhancing applications.
paul.verschure@gmail.com

Riccardo Manzotti

Preface

The notion of aesthetics revolves around the notion of the subject. This statement does not necessarily imply that the fundamental concepts of aesthetics have to be flattened merely to those of psychology. Neither does it imply that aesthetics has to be reduced to a subfield either of psychology or of neuroscience. It is even questionable whether psychology will ever be reduced to a subset of neurosciences (Kenny 1984; Faux 2002; Manzotti and Moderato forthcoming).

Indeed the notion of the subject is definitely larger than the current scope either of psychology and *a fortiori* of neuroscience. Our understanding of the subject and, inevitably, of the relation between subject and object is a keystone of our metaphysical understanding of the world. It is impossible to advance fruitfully in any philosophical discussion as to the nature of the world without some implicit or explicit intuition about the subject.

After all, any theory of aesthetics rests on our understanding of what a subject is (Levinson 2003). Since Descartes, the subject has been identified with the mind and its underpinnings. Even the analytical oriented philosophers of aesthetics, who deny that we could ever gain any better understanding of aesthetics from psychological data (Dickie 1962; Currie 2003), ground their views on something akin to Wittgenstein's model of the mind. This dependence partially explains the emphasis that aesthetics tends to give to the importance of logic and language. Yet in the last twenty years, largely due to the advances in neuroscience, there has been an intense debate as to the nature of the mind.

In other words, claiming that aesthetics resides in the subject is not an attempt to resurrect the debate as to whether aesthetics is based on the mythical existence of a special aesthetic experience, that will one day be identified and dissected by either psychology or neurosciences (Beardsley 1961; Dickie 1961; Dickie 1965; Beardsley 1969). A recurring debate seems to be polarized by two opposite views. On the one hand it is maintained that 'Psychology is not relevant to aesthetics' (Dickie 1962, p. 285). On the other, it is suggested that the roots of any artistic process is to be found in neural processes. Consider Semir Zeki's claim that 'There can be no satisfactory theory of aesthetics that is not neurobiologically based' (Zeki 2001, p. 52). We don't want to resolve this debate, nor are we looking for aesthetics inside the subject's psychophysical and neural machinery; rather we are trying to flesh out a very broad framework encompassing epistemological, phenomenal and ontological issues. Once this framework is established, it will perhaps become possible to conceive aesthetics from a different perspective. For standing back from the specifics of the arguments referred to above, it would, on reflection, be extraordinary if one could redefine the ontology of the subject without any side effect on aesthetics.

It is fair to maintain that in recent years there has been a growing tension as to the nature of the subject. On the one hand many authors, heralding the neuroscientific view, have suggested that the subject is reducible to neural activity. On the other hand, some scholars suggested considering more radical solutions and frameworks that, although remaining inside the physicalist playground, could nevertheless encompass a wider set of physical phenomena. If the former solution were adopted, and the subject turned out to be nothing but neural activity, then some kind of neural-centred view of most, if not all, human activities should eventually prevail. However, in the light of present knowledge, this may not necessarily be the case.

The alternative, which we consider and outline below, is to relocate the subject, freeing it from the narrow walls of the cranium and allowing it to encompass and swallow up the surrounding perceived and conceptualized environment. To date this option has been dubbed externalism and has been articulated and differentiated in many intertwined branches: semantic externalism, phenomenal externalism, cognitive externalism, vehicle externalism, and many others. Several of these approaches will be outlined in subsequent chapters in this volume.

For our purposes we shall take externalism in a broad sense, i.e. to mean any view that considers some portion of the physical environment surrounding the subject's body as somehow constitutive and necessary to the occurrence of the kind of mind at hand. For instance, if you are a cognitivist focusing on cognitive processes and if you believe that handling a tool is constitutive of your cognitive processes, then in our view you are an externalist (Clark and Chalmers 1998; Clark 2008). Or, if you happen to believe that the essential composition of water is relevant as to your semantic content, you are an externalist too, albeit of a different kind than the former (Putnam 1973). Yet, both views, as different as they are, share the belief that a certain aspect of the mind is constituted by events, processes, or circumstances involving the world external to the body of the subject. The kind of externalism that will be considered by most of the authors in this volume is more radical and ontologically more committing than either semantic externalism or simple embodiment.

Preface

In this book we consider various versions of externalism. By and large, externalism is the view that the external world is relevant and indeed constitutive of the subject, which is more extended than the body. In particular, externalism is taken as the view that the physical underpinnings of the mind are spatio-temporally more extended than the neural activity inside the nervous system. For the purposes of this volume, the key is the fact that a shift in the subject's ontology will inevitably have repercussions for any theory of aesthetics.

If one had any doubt as to the dependence of the model of aesthetic experience on the subject's ontology, one example will suffice. Consider the notion of representation and how much it depends on the way in which the subject refers to and, of course, believes to represent the external world. It is not fortuitous that most discussion as to the nature of representation has been kidnapped by the philosophers of mind. In this area the impact of externalism could hardly be greater, since in most of its variants it involves a fundamentally anti-classical-representationalist stance. If there are no 'representations' in the mind, what can we say of representation in art?

At the onset, it is important to stress that in this volume we try to avoid any form of reductionism or simplified form of psychologism. Our aim is to highlight a developing, perhaps still to be perfected, externalist view of the subject and then to see how such a view would affect aesthetics and artistic experience.

If the subject were extended to a spatio-temporal collection of processes and properties more extended than the body, how would this change our understanding of what art is? The crucial question posed by aesthetics is about the value, nature and conditions of art. But art does not live in a void. Art's natural environment is made by subjects and their relations among themselves and with the world. A relocated and extended subject will inevitably lead to a different notion of art and thus of aesthetics.

If externalism were correct, aesthetics would have to change accordingly to cope with an extended subject — which is to say that a situated aesthetics would be needed. The goal of the following papers is to outline such an extended notion of aesthetics which no longer conceives art as something that has to elicit or trigger responses in human brains, but rather as a network of spatio-temporal processes and perceived relations between individual and worldly matters and events.

Situated aesthetics is an empirical endeavour as well as a conceptual challenge. On the one hand it suggests that there is a physical and perceptual foundation to aesthetic experience which is not constrained by the boundaries of the nervous system. At the very least, it would reveal the spatio-temporal physical and perceived playground in which aesthetic experience can take place. There is no reason why such an approach should diminish or discard the important findings of neurosciences. On the contrary, situated aesthetics is going to take advantage of them, while going beyond to encompass a broader network of processes.

An important caveat as to the name chosen for this book: *situated aesthetics*. It does not imply a close association with the notion of situated cognition as outlined elsewhere (for instance in Robbins and Aydede 2009). By situated aesthetics we mean an attempt to develop a theory of aesthetics based on the very broadly conceived hypothesis that the relevant facts, processes and properties constituting a mind are not confined to the boundary of the nervous system — that the mind is larger than the body. Of course, there are still many notions as to what the mind is and many conflicting aspects and approaches that remain unresolved. But essentially there are two fundamental intuitions at stake: either the mind is made up of neurons the same way mechanical force is produced by muscles even though in a much more complex manner, or the supervenience basis of the mind somehow extends beyond the body boundaries.

As with the term 'externalism', we are going to use 'situated aesthetics' in a similarly broad and flexible way. Whenever the aesthetics under discussion requires underlying externalist ontology, it will be considered a case of situated aesthetics. At the same time it is not to be confused with other etymologically akin labels such as environmental aesthetics (Carlson 1992; Fisher 2003). Situated aesthetics is neither aesthetics about the environment nor aesthetics of environmental art. Situated aesthetics is aesthetics dependent on the adoption of externalist ontology of mind.

It's too early to give an exact definition of what is meant by the adjective 'situated'. In the cognitive sciences, situated cognition seems to focus on the cognitive aspects of the mind — something not too far from Clark and Chalmers' model of the extended mind (Clark and Chalmers 1998; Clark 2008). It is a definition that seems to set aside the issue of phenomenal experience. Here 'situated' does not necessarily have such a restricted meaning. Some of the authors in the following chapters explicitly consider a full extension of the mind — thereby considering phenomenal experience as well — while others restrict their approach to more specific aspects of the mind.

For our purposes, what is relevant is (i) considering a theory of aesthetics based on an externalist model of the mind, and (ii) using a theory of the mind whose physical and perceptual underpinnings are not limited to activity in the nervous system. As long as both requirements are met, the outcome is taken to be a case of situated aesthetics, at least as far as this book is concerned.

The book is divided into three parts. In the first the authors describe, analyse and even criticize different forms of externalism and highlight their consequences for various aspects of aesthetic experience. In the second part, situated aesthetics is confronted with different artistic forms.

Finally, in the third and last part, some aspects of situated aesthetics are grounded in the aesthetic experiences provided by specific artworks.

At the onset, it is mandatory to map the territory all the more because of the confusion among different authors as to the use of terms like situatedness, externalism, and enactivism. For this reason, **Riccardo Manzotti** offers a review of current externalist approaches in neuroscience, cognitive science and philosophy of mind. Externalism is by no means a simple view and many authors differ widely in their assumptions and their use of the relevant terms. In order to frame a reference point, Manzotti addresses the 'location question' as to where the mind is physically located. Since different authors focus on different notions of the mind, it follows that there are at least as many versions of externalism as definitions of the mind. Confronting internalism and externalism, it is thus possible to define a taxonomy of externalist views so as to sketch the theoretical landscape of externalism. Eventually, Manzotti wonders what the consequences are for aesthetics.

Erik Myin and **Johan Veldeman** unfold an externalist approach by outlining the distinction between active externalism and explanatory externalism. Their goal is twofold: first, they analyse the pros and cons of various forms of externalism as to cognitive mental processes; secondly, they apply the outlined theoretical landscape to a selection of artistic cases in aesthetics. They begin considering the so-called parity principle that takes the functional equivalence between external and internal cognitive processes as the basis for the claim that the mind can extend into the world. Yet, the parity principle is not satisfactory since, in the most interesting cases, the external world when interacted with by the agent is indeed constitutive of processes which could not take place without such interaction. In this sense, externalism suggests that environment is not only a substitute of internal processes but rather the source of new mental processes. This strong point is put to good use in the second part of Myin and Veldeman's work where they show how aesthetic theories have been biased by internalist views of the subject. Moreover, the traditional discussion about the role of aesthetic experience, as exemplified by the contrast between Beardsley and Danto, gets a twist from an externalist perspective in so far as the traditional perceptual chain is substituted by an enacted loop between subject and work of art. Myin and Veldeman's longshot is whether a broadly enactivist approach allows the domain of aesthetic experience to advance beyond the Cartesian limitations of an internalist phenomenology. Aesthetic experience is reframed as an activity that encompasses some of the epistemic, social and historical aspects stressed by many recent scholars of aesthetics (from Danto to Walton).

As to externalism, the debate on perception and phenomenal experience is often geared towards vision and touch. This is a misleading opinion perhaps caused by a somehow more passive notion of the other sensory modalities like hearing, smell, or taste. As to hearing, **Joel Krueger** challenges this misconception, outlining the beginning of an enactive account of auditory experience—particularly the experience of listening sensitively to music. Thanks to an externalist perspective, music is no longer cramped into the realm of pattern recognition. Rather it can be conceived as an active skill that involves a physical interaction with the space where the music is heard and performed. Listening to music is not only a process of cortical processing of frequencies, it involves either covert or actual motor involvement. Taking advantage of a sensorimotor oriented version of externalism—namely enactivism—Krueger investigates how sensorimotor regularities grant perceptual access to music *qua* music. Thus phenomenal musical experience may unfold a complex spatial content and, by means of sensorimotor regularities, it is possible to foresee its content. The notion of musical space is exploited so as to admit both an individual and a social dimension of music perception.

At the onset it is clear that the notion of space is of paramount importance for the externalist endeavour. The space is the domain where the subject perceives, extends and develops. However, we are not referring to a simplified notion of strictly geometrical volume, because the space of perception is highly subjective. Space has to be reconceived in terms of causal relations, sensorimotor regularities, physical processes, and subjective operations. Aptly, dwelling on the important and sometimes overlooked phenomenological tradition, **Liliana Albertazzi** focuses on the structure and nature of extended space whose relevance is paramount both for the understanding of mind and for any forthcoming aesthetics. The 'extended space' is at the same time a structure of our aesthetic experience and of the perceived physical world. Thus, the extended space is neither a purely phenomenological description of the lived nor a merely physical notion, but rather a concept that could be used as an explicative bridge between externalist and internalist views, as well as phenomenological and more physical oriented perspectives. Such a notion, which stems out of a revision of the traditional internal/external dichotomy, is then put to service in the case of pictorial representation. Thus Albertazzi, whose articulated position does not easily overlap with simplistic forms of physicalistic externalism, is not an internalist in the classical AI and neuroscientific perspectives. On the contrary, she shows that the way in which the subject is embodied in the environment is indeed constitutive of the work of art.

In the second part, the idea of externalism is brought to bear on a specific branch of aesthetics. Various art forms are explicitly addressed.

Robert Pepperell opens the second part showing how some artists and art theorists have understood aesthetic activity as a distributed phenomenon, extending beyond any individual person or mind. This is at variance with the view of aesthetic experience implicit in neuroaesthetics, which seeks to account for art in terms of localized brain activity. As most of the defenders of externalism, Pepperell adopts his personal version, dubbing it extensionism, which is a perspective he exploits to look at a work of art. Extensionism stresses the extended dimensions of objects and events rather than the distinctions between them. When this approach is applied to the analysis of art it reveals the widely distributed nature of artworks and the mental qualities they convey. This is correlated with a view of the mind that extends far beyond the head. Art is no longer a means to trigger the appropriate aesthetic experience. Art is beyond the skin.

If aesthetics can be extended to tools, works of art, ways of handling objects, in one of the most thought provoking and original papers, **Lambros Malafouris** wonders how to probe into the cognitive past of humans. By means of a situated approach to aesthetics it is conceivable to overcome the apparently impassable chasm dividing our phenomenal world and that of past humans. Malafouris advocates the collapse of the dividing lines between perception, cognition and action. The rejection of the methodological separation between aesthetic experience and embodiment paves the way to a different way to reconstruct the mind of humans. Archeology and externalism can thus provide together a new understanding of phenomenal experience. Yet, the authors warns against easy enthusiasms. Although a situated aesthetic approach might offer the opportunity to explore alternative forms of aesthetic experience and ways of seeing, a more appropriate ontological foundation against which to place aesthetics is indeed mandatory. Malafouris considers in some detail the concrete example of pottery in archeology. By the careful analysis of pottery, it is possible to reconstruct the complex entanglement between the potter's sensing and the process of perceiving and fleshing out of the vessel. Sometimes it is sensed where the hand meets the surface of the clay and sometimes where clay meets the potter's eye. In other words, archeology can venture to single out the process of active material engagement that is at the root of aesthetic experience.

As promising as it may seem, situated aesthetics runs the risk of being only a fashionable label. It is thus proper to consider critically the usage and the shortcomings of such a term. **Sabine Marienberg** accomplishes thus a twofold goal in her paper: on the one hand, she criticizes some simplified use of the notion of 'situatedness' by and large; on the other hand, she puts to the test the notion of situatedness, addressing the issue of language and rhythm. Stating explicitly that she does not have blind faith in situated aesthetics, Marienberg suggests an acid test: is the new approach able to discover something so far unknown about linguistic experience? Inspired by an unabashed Peircian perspective, she applies his critical method to poetic speech with special attention to the materiality of language. Are the sensible and phenomenal properties of language—paradigmatically expressed by rhythm and musical properties—separable from the situation in which language is learned, generated, and perceived? So poetry is presented as a case in which the materiality and situatedness of language is a non-negotiable aspect. And rhythm is an expression of the twofold nature of speech where the corporeality of the voice and language rub against each other like on an erotically charged contact surface.

From language to literature the distance is indeed not so great. **Paola Carbone** exploits her knowledge of contemporary literature to see how much the explicit awareness of situatedness oriented and influenced the work of the main writers in English literature. In particular, she tries to make as clear as possible the importance of embodiment and situatedness in the developments in the coming to life of literary works through the 'great labour' of young poets and through individual reading experiences and actual interactions with existing worlds. From an externalist point of view, one can say that a narration is neither subjective nor objective, but rather emerges out of the actual interactions constituting both the subject and the world. Manifestly, crossing time, space and cultures, literature allows us to single out affinities as to the way in which individuals perceive the world. Storytelling embodies an enaction of our mind, and, as postmodernism underlined, the reader is the author of the literary work as much as the author is its first reader. Through the fascinating literary works of the eighteenth century up to contemporary digital literature, Carbone leads us to discover the constitutive role of the author's physical and cognitive world.

Finally, in the third part, we consider individual artists whose work is explicitly oriented to or inspired by the idea of an extended subject.

No author merges together more happily the practice of philosophical theorizing with the joy of artistic creation than **Teed Rockwell**. Taking advantage of first-person experience as a musician and moving from his personal brand of externalism, he deals with an unanswered question as to the nature of musical experience. He begins by distinguishing between phenomenal experience by and large and aesthetic experience. Although a long established tradition considers the former closer to perception and the latter closer to inner thoughts and feelings, this is not necessarily the case. In fact, both cases of experience show a strong dependence on the actual external world. In particular he focuses on Leonard Bernstein's hypothesis as to the Chomskian structure of musical space. Bernstein creates a musical analogy with Chomsky's concepts by seeing the diatonic scale as roughly analogous to the phonemes and morphemes of spoken language. Rockwell discusses at length such an analogy and shows how inadequate it is. On the contrary, he maintains that we can be aware of what music means without being able to analyse it into parts, and because the meaning resides in the music itself, rather than being tacked on afterward by our minds. The music thus is not in the head.

Art is evolving and new media and devices are changing the artistic media in unexpected ways. The possibility offered by new technologies to shape a new interactive space where the subject is challenged by a new reality is a very timely topic. **Sylvain Le Groux** and **Paul F.M.J. Verschure** outline a new exciting theoretical and technological field of situated musical composition. With the advent of new interactive technologies, computer-based music systems evolved from sequencers to algorithmic composers, to complex interactive systems. Consequently, the frontiers between composers, computers and autonomous creative systems have become more and more blurry, and the concepts of musical composition and creativity are being put into a new perspective. Dwelling on real interactive setups they show how the creation of art can be rekindled by new approaches exploiting the situated nature of musical experience. They maintain that in order to improve the understanding of compositional processes and to design machines capable of supporting expressivity and creativity of musical composition, an externalist notion of aesthetics is indeed needed. Eventually, they outline a situated and interactive approach to music composition, and present a real case of situated computer-generated music.

The next chapter shows a series of real cases of specific works of art explicitly trying to blur and dissolve the separation between the inner self and the outer world. **Stéphane Dumas** describes and discusses examples of artworks sharing a common feature: the conceptual and sometimes concrete dissolution of the boundaries. This is obtained by means of new technologies ranging from advanced examples of bioengineering to biotechnologies. The paper can thus express a model inspired by cutaneous secretion — as is done by live skin — so this model establishes a relationship between language and secretion. Again Dumas shows that aesthetic experience does not take place inside neural processes. In particular two artworks are considered: Paula Gaetano Adi's Alexitimia, a robotic sculpture dealing with skin, and a robotic artist (MEART), hybridizing biological and artificial neural networks. Eventually, Dumas proposes a critique of the notion of representation, establishing a link between different theories of embodiment and the aesthetic model of creation as secretion. Cognition is no longer based on the representation of a world by a subject processing information in order to build a map of that world. The secretory model functions more like a contamination, a dissemination and an interference than a projective representation: it works by hybridizing rather than by purifying.

In the final chapter, **Giuliano Galletta** outlines one more example of the inadequacy of traditional notions of self and inner experience. He adopts an autobiographical perspective based on his work as an artist. During his life he kept building an artistic alter-ego of himself, a fictitious but visible 'Giuliano Galletta' that like himself collected memories, developed tastes and goals. In his artwork, Giuliano Galletta blurs the separation between his artwork, his self, and the beholder of his production. Similarly, in his paper, he does not present a linear description of his artistic endeavour; he prefers to wander seamlessly and shamelessly among various inspiring factors that contributed to the construction and public growth of that fictitious self which is now everywhere but inside him.

All in all, this volume aims at outlining the beginning of a new approach to the mind and to aesthetic endeavour. Whether this rather pristine and somewhat unavoidably naïve perspective will be successful or not, only the future will tell. However, it is fair to maintain that the scholars in this volume believe that externalism, in some of its variants, captures a mandatory aspect of the mind. If there is any merit in this view, aesthetics cannot be neutral and thus needs to be reconceived. All in all, this collection of papers offers an interesting cross-section of externalist approaches to aesthetics, ranging from a more theoretical analysis up to a first-person artistic experience. The contributions are for this reason written at various levels of abstraction and with different degrees of interdisciplinarity. I am aware that the aim of the volume — namely fleshing out an externalist aesthetic perspective — could fairly seem a double longshot to many. Not only is externalism still to be an established view, but many will argue against any dependence between a theory of the mind and aesthetics. Whether or not these counter-arguments are cogent, there does seem to have been a hubris-like overconfidence among two opposite camps: on the one hand, hard-core philosophers rejecting any relation between the mind and aesthetics; on the other hand, brain scientists confident of the possibility to reduce aesthetics to a subfield of neuroscience. This book aspires to steer between these two extremes trusting that aesthetics is a mandatory test for any theory of mind and, at the same time, that we cannot understand art without understanding the mind.

References

Beardsley, M. (1961), 'The Definitions of the Arts', *The Journal of Aesthetics and Art Criticism*, **20** (2): 175–87.
Beardsley, M. (1969), 'Aesthetic Experience Regained', *The Journal of Aeshtetics and Art Criticism*, **28**: 3–11.
Carlson, A. (1992), 'Environmental Aesthetics' in D.E. Cooper, Ed., *A Companion to Aesthetics*, Oxford, Blackwell: 142–4.
Clark, A. (2008), *Supersizing the Mind*, Oxford, Oxford University Press.
Clark, A. and D.J. Chalmers (1998), 'The Extended Mind', *Analysis*, **58**: 10–23.
Currie, G. (2003), 'Aesthetics and Cognitive Science' in J. Levinson, Ed., *The Oxford Handbook on Aesthetics*, Oxford, Oxford University Press: 706–21.
Dickie, G. (1961), 'Bullough and the Concept of Psychical Distance', *Philosophy and Phenomenological Research*, **22** (2): 233–8.
Dickie, G. (1962), 'Is Psychology Relevant to Aesthetics?', *The Philosophical Review*, **71** (3): 285–302.
Dickie, G. (1965), 'Beardsley's Phantom Aesthetic Experience', *The Journal of Philosophy*, **62** (5): 129–36.

Faux, S.F. (2002), 'Cognitive Neuroscience from a Behavioral Perspective: A Critique of Chasing Ghosts with Geiger Counters', *Behavior Analyst*, **25**: 161–73.

Fisher, J.A. (2003), 'Environmental Aesthetics' in J. Levinson, Ed., *The Oxford Handbook of Aesthetics*, Oxford, Oxford University Press: 667–78.

Kenny, A. (1984), 'The Homunculus Fallacy' in A. Kenny, *The Legacy of Wittgenstein*, Oxford, Basil Blackwell.

Levinson, J., Ed. (2003), *The Oxford Handbook of Aesthetics*, New York, Oxford University Press.

Manzotti, R. and P. Moderato (forthcoming), 'Is Neuroscience the Forthcoming "Mindscience"?', *Behaviour and Philosophy*.

Putnam, H. (1973), 'Meaning and Reference', *The Journal of Philosophy*, **70**: 699–711.

Robbins, P. and M. Aydede, Eds. (2009), *The Cambridge Handbook of Situated Cognition*, Cambridge, Cambridge University Press.

Stevens, W. (1971), *The Collected Poems*, New York, Knopf, Alfred A.

Zeki, S. (2001), 'Artistic creativity and the brain', *Science*, **293**: 51–2.

Part One
From Externalism to Aesthetics

Riccardo Manzotti

Varieties of Externalism and Aesthetics

> *There is a long history of discussions of the aesthetic and of art in which the fundamental concepts are psychological, in the sense of being or including concepts of states of mind.*
>
> Iseminger (2005, p. 99)

Since its alleged beginning in the eighteenth century, aesthetics has been deeply interwoven with the assumed model of the subject. It is not by chance that the Kantian notion of aesthetics stemmed out after he developed his model of the transcendental subject. After all, aesthetics is the branch of philosophy devoted to conceptual and theoretical enquiry into art and aesthetic experience (Levinson 2005, p. 3). Since art is made by human subjects and aesthetic experience is an important part of subjects' life, it's natural to conceive aesthetics and philosophy of art inside the playground set by the accepted notion of the subject's nature. In this respect, if philosophy of art is not neutral as to what the subject is, it is mandatory to render explicit the ontology at hand. This is even truer considering that most conceptions of art are either concerned with perception (Kant, Clive Bell, Roger Fry) or with expression (Benedetto Croce, Lev Tolstoy) or with phenomenal/semiotic representation of the external world (George Dickie, Beardlsey, Arthur Danto, Kendal Walton). All these aspects are based on the accepted notion of the subject's nature.

In the last couple of decades, our understanding of the nature of the subject underwent heavy strain because of the dramatically increased amount of neurological data on conscious human subjects. As a result, many scholars consider it feasible to outline a neurological model of the mind. On the one hand, neuroscientists are looking for the neural underpinnings of most if not all of our mental states (Dehaene, Changeux *et al.* 2006; Tononi and Koch 2008; Koch 2010; Revonsuo 2010). On the other hand, both aesthetic experience and aesthetics in general are not exempted, as is shown by the efforts of many neuroscientists (Ramachandran and Hirstein 1999; Zeki 2000; 2001a; 2002; 2009; Calvo-Merino, Jola *et al.* 2008). Although it cannot be denied that neuronal data offer an impressive overview of the underlying neural mechanisms, it is far from being obvious whether everything mental is reducible to what the neurons do. Many mental aspects seem to offer different reasons to resist such a reduction, notwithstanding the unabashed enthusiasm of many neuroscientists.

For example, consider Cristoph Koch's certainty that 'Scientists are now revealing the material basis of the conscious mind. In coming years they will gradually fill in the details, making much of the armchair philosophizing moot... Such theories will provide quantitative answers to questions that have long stumped us' (Koch 2010, p. 76). For one, there are not yet any available theories explaining how and why quantitative phenomena should lead to the emergence of qualitative phenomenal qualities. Nor has it been explained in the least how to connect the normative element of many mental phenomena to the physical domain. In this regard, Shaun Gallagher stresses that 'neither the cognitive neurosciences nor phenomenological approaches to consciousness, however, should be satisfied with simple *correlations* that might be established between brain processes described from a third-person perspective and phenomenal experience described from a first-person perspective. Such correlations do not constitute *explanations*, and indeed, such correlations are in part what need to be explained' (Gallagher 2005, p. 6). And, of course, on top of such epistemic concerns as to the possibility to bridge the gap between correlations and explanations, there are even stronger concerns as to whether knowing more about how the brain performs certain mental occurrences will ever improve our understanding of the implemented epistemologically higher functions (Currie 2003). Notwithstanding the great success in many respects of neuroscience, many scholars do not share such faith in neuroscience as the forthcoming 'mindscience' (for a review see Manzotti and Moderato forthcoming). There are aspects of the mind that do not match our current understanding of the nature of the physical and *a fortiori* of the neural domain. As a result of these and similar issues, a few authors raise some scepticism as to the soundness of such neural chauvinism (Faux 2002; Bennett and Hacker 2003; Noë and Thompson 2004; Manzotti and Moderato forthcoming).

This sketchy introduction ought to stress a rather trivial point—namely that the reduction of the mental to the neural is far from being accomplished and it is not even necessarily going to ever be so. If this is true, it means that it is worth considering alternative options as to the physical underpinnings of mental phenomena. These options should not only add new hypotheses, but rather redefine the conceptual terrain and the network of premises commonly assumed. In short, if the mind would depend on something else than neural firings, in turn aesthetic experience would too.

Among the various alternatives, in this book we are considering the externalist hypothesis—could the mind depend and indeed be physically constituted by a nexus of processes larger than the brain? The answers to this question not only provide important insights into the nature of experience; they also tell us something about the nature of aesthetic experience and the nature of art itself. Therefore, the goal of this chapter is to outline the various views competing to offer a physical foundation for the human mind. It is very important to understand how deep the difference is between their ontological roots and how far such differences are going to reverberate in the corresponding aesthetic theories.

1 The location question

As to the mind, it is often assumed that the most basic question is about the nature of the mind. What is the mind? What is the stuff the mind is made of? If you are a physicalist, as I assume you probably are, you should conceive the mind as something physical (in this I agree with most available scientific and philosophical literature; see Dennett 1995; Kim 1998; Strawson 2005). This is a very broad statement that is not particularly committing and yet it is sufficient to draw some preliminary basic suggestions. If the mind is a physical phenomenon, the mind has to be located somewhere in space and time. This means that it should be possible to pinpoint a spatio-temporal region that corresponds to a certain mental occurrence. If we focus on the phenomenal side of the mind, as indeed I am going to do in this chapter, the same will hold—in principle it should be possible to pinpoint a spatio-temporal region corresponding to a certain mental occurrence. Where is the mind as a physical phenomenon? And when does the mind occur? Let us call the combination of these two questions the *location question*.

Looking for a spatio-temporal region is perhaps a convoluted way to look for what is commonly referred to as a process. This is not a small conceptual step, shifting from other widespread and less physically-sound yet philosophically popular notions such as states of affair, states, or events. For in physics what happens is usually closer to the occurrence of a process rather than to the instantiation of a state of affairs. Just to give a few examples, in physics it makes more sense to speak of the process of oxidation rather than of a state of affairs corresponding to oxidation. In other words, the elementary building blocks of physical reality are not static entities but rather processes taking place in space and time (Whitehead 1925; Eddington 1929; Heisenberg 1958; Pylkannen 2007). But even without committing too strictly to a process-ontology, it is rather uncontroversial that any physical phenomenon could be located in time and space. If phenomenal experience is a physical phenomenon—and I do not see any viable alternative—it ought not be an exception.

The location question lies at the very foundation of current scientific research on the nature of subject and so far there is neither consensus nor hope of getting closer to a solution anytime soon. Yet it is possible to envisage two broad rivalling alternative views whose respective value is still largely unresolved.

On the one hand, many authors tried to locate the relevant physical phenomena inside the physical boundary of the subject's body — often inside the physical limits of the nervous system or of some subset of it. This trend can be labelled *internalism* — namely, the idea that the mind depends on or is identical to properties or events taking place inside the boundary of the subject's body or some subset of it. Internalism is the view that all the conditions that constitute a person's thoughts and sensations are internal to the nervous system (Koch 2004; Adams and Aizawa 2008; Mendola 2008). Consider Koch's claim that 'If there is one thing that scientists are reasonably sure of, it is that brain activity is both necessary and sufficient for biological sentience' (Koch 2004, p. 9). The crucial claim in this sentence is whether brain activity is *sufficient* or not. Internalism advocates that it is, while other views either weaken to a certain extent or straightforwardly reject the sufficiency of brain activity. As for the necessity, more or less all scientists and most philosophers of mind alike agree, although what kind of necessity is still an open issue. Coherently, internalism can be seen as the view that answers to the location question by pinpointing to some neural process. According to internalism, the spatio-temporal boundaries of mental phenomena are limited to a short span of a few hundreds of milliseconds and to the bursts of neural firings spreading through axons, dendrites and cellular bodies. An aesthetic experience would allegedly correspond to some neural process initiating and ending inside the nervous system although undoubtedly originated by external stimulation, learning, and development.

On the other hand, many scholars are sceptical as to the resources of the body alone. In particular, it doesn't seem plausible that the internal properties of a body might cope with certain aspects of the mind whose properties seem unmatched by the properties shown by neural activity. Among such resistant features of the mind, two broad categories can be outlined: semantic/intentional/relational properties and phenomenal/first-person properties. The first group expresses the fact that the mind seems to be projected outward. It seems to proceed from the body towards events scattered in time and space accordingly to the subject's ends and stimuli. The mind seems to have a non-reducible relational and externally oriented attitude expressed by hard-core philosophical issues such as intentionality and semantics. As a result, it has been considered whether the environment and the spatio-temporal physical surrounding of the subject's body could be literally constitutive of the mind. This other approach has been labelled *externalism* — the view that the mind depends on or is identical to properties or events taking place outside the boundary of the subject's body. In its more radical form, as we will see, an externalist answers to the location question by suggesting that the spatio-temporal boundaries of the physical processes identical to the mind are, at least in principle, larger than those of the body and temporally as extended as they need to be to encompass what they refer to (in practice a lot more than a few hundreds of milliseconds).

From the onset it is worth stressing that externalism is no less physicalist than internalism. This should not come as a surprise although it often does. There is a widespread and rather surprising misconception according to which only internalism is a genuine physicalist position. As a matter of fact, physicalism requires only that the mind is explained in physical terms. Physics is not made by the nervous system alone — being 'neural' is a subset of being 'physical'. Thus, by and large, there are many physicalist explanations which are not confined to the neural domain. Equating the mind with the brain is not the only possible option for physicalism. However, it is the only possible option for a physicalist who embraces internalism in some form. Yet, for the physicalist, internalism is only one possibility among many and there is no conclusive evidence for it.

In the following, after having outlined the present state of the internalism vs. externalism debate, the various forms of externalism will be listed. The variety of the available models of externalism highlights the value of externalism as a broad approach. In the near future, it is probable that some of these versions will be discarded — some because they are too daring and others because they are too conservative. It is something to be expected.

2 Internalism vs. externalism

Notwithstanding the current interest in embodiment, situated cognition, ecological perception, and externalism in their various versions (Finlay, Darlington *et al.* 2001; Thompson and Varela 2001; Hirose 2002; Shanahan 2005; Pfeifer, Lungarella *et al.* 2007; Rakic 2009; Robbins and Aydede 2009), many authors assume that an isolated healthy and working brain is all you need to get mental content.

Of course, a healthy brain requires development, feedback, sensorimotor integration, embodiment, and a surrounding environment. Yet, once the brain is ready with all the neural connections in the right place, the pivotal issue is whether it is *sufficient* for the occurrence of a mind. As we have seen either explicitly as in Koch's words or implicitly as in the premises of many neuroscientists' works, the brain is held sufficient to produce conscious experience. As a corollary or as a proof depending on the chosen line of reasoning, it is held that, in principle, the brain could be disconnected from the environment and it would continue to be sufficient for the emergence of awareness, phenomenal experience, will, and all kinds of mental content. Thus it would become the mythical brain in a vat. Or, at least, many believe so.

The belief that, once developed, the brain can work in isolation and produce phenomenal experience is supported by the internalist stance and *vice versa*. Internalism is the view that all the conditions that constitute a person's thoughts and sensations are internal to the nervous system (Goldman 1999; Goldberg 2007; Adams and Aizawa 2008; Mendola 2008). Yet, internalism is only a possibility and so far there is not any conclusive evidence for it.

Among neuroscientists, internalism is the default view to the extent that many do not feel the need to explicitly mention it as a premise of their work. As an example, consider pain in phantom limb patients. Often, patients who have lost a limb report having sensations from the missing limb. Since the limb is no longer there and yet it is felt by the subject, the prevalent interpretation consists in supposing that the brain is concocting the feeling of the severed limb. Once again it seems that the brain can produce phenomenal experience by itself. In particular many of these feelings consist in painful feelings to various degrees. To explain such pain in the absence of a corresponding damaged part of the body, Ronald Melzack developed the neuromatrix theory of pain proposing that pain is produced by 'neurosignature' patterns of nerve impulses generated inside the nervous system by dedicated neural networks that all together constitute the 'body-self neuromatrix' in the brain. Apart from a rather striking resemblance to Muller's long abandoned theory of specific energies, the key hypothesis is that although these neurosignature patterns may be triggered by sensory inputs, they may also be generated independently of them. Therefore their mental aspect is independent of the external cause. The negative mental quality of pain is taken to be the result of some intrinsic and internally generated property of neural patterns. Melzack thus concludes that

> All the qualities we normally feel from the body, including pain, are also felt in the absence of inputs from the body; from this we may conclude that the origins of the patterns that underline *the qualities of experience lie in neural networks in the brain*; stimuli may *trigger the patterns but do not produce them*. (Melzack 1993, p. 619, italics mine)

Once more it is assumed that what we call phenomenal qualities arise out of the neural patterns of activity generated inside the brain. In short, according to Melzack, the content of our experience is produced internally in the brain — the qualities of experiences are underlined by patterns occurring in the cortical neural network.

Other neuroscientists defend similar views — a few quotations will be enough.

According to Anil Seth, 'Any scientific study of consciousness is based on the premise that phenomenal experience is entailed by neuronal activity in the brain' (Seth, Izhikevich *et al.* 2006, p. 10799). Along the same line, Jessie Prinz stated that it is 'a central plank of modern materialism—the supposition that consciousness supervenes on the brain. I have yet to encounter an argument that is nearly powerful enough to consider giving up the brain doctrine' (Prinz 2009a, p. 425). Antti Revonsuo recently argued that 'The mainstream empirical theories in cognitive neuroscience all seem to accept internalism, at least implicitly' (Revonsuo 2010, p. 222). As already mentioned, Cristof Koch unabashedly stated that 'the entire brain is sufficient for consciousness—it determines conscious sensations day in and day out... [It is] likely a subset of brainmatter will do' (Koch 2004, p. 87) and 'If there is one thing that scientists are reasonably sure of, it is that brain activity is both necessary and *sufficient* for biological sentience' (Koch 2004, p. 9, italics mine). All in all, most of the leading neuroscientists in the field of consciousness accept the hypothesis of the sufficiency of the brain as to the emergence of the mind (Crick and Koch 1998; Llinàs 2001; Lamme 2003; Zeki 2003; Koch 2004; Tononi 2004; Dehaene, Changeux *et al.* 2006; Lamme 2006; Tononi 2008), something close to what Alva Noë and Evan Thompson dubbed as the thesis of the *minimally sufficient neural substrate*—that is 'the commitment that for every conscious state there is a minimal neural substrate that is nomically sufficient (as a matter of natural law) for its occurrence' (Noë and Thompson 2004, p. 5).

Only a few authors inside the internalist trench raised some scepticism. Most notably Ned Block affirmed that he never heard anyone stating 'that if a fusiform face area were kept alive in a bottle, the activation of it would determine face-experience—or any experience at all' (Block 2007, p. 482). Yet in the debate between enactivism and internalism he opposed the former. Further, internalism faces strong criticism from the many defenders of various forms of enactivism, externalism, and situated cognition (Varela, Thompson *et al.* 1991/1993; Dretske 1995; Dretske 1996; Lycan 2001; Rowlands 2003; Tonneau 2004; Rockwell 2005; Byrne and Tye 2006; Honderich 2006; Hurley 2006; Manzotti 2006d; Manzotti 2006a; Manzotti 2006b).

The internalist view is adopted, understandably, by most neuroscientists but, surprisingly, also by many if not most philosophers. Consider Jaegwon Kim's claim that

> if you are a physicalist of any stripe, as most of us are, you would likely believe in the local supervenience of qualia—that is, qualia are supervenient on the internal physical/biological states of the subject. (Kim 1995, p. 160)

Physicalism is equated with believing that the mind stems only out of activity internal to the nervous system (local supervenience on the internal physical/biological states of the subject). On a rather similar note, a philosopher like Jerry Fodor stated that 'mind/brain supervenience(/identity) is our only plausible account of how mental states could have the causal powers that they do have' (Fodor 1987, p. 44). When Patricia Churchland states that 'I am a materialist and hence believe that the mind is the brain' (Churchland 1985, p. ix), there seems to be no doubt about what she means. All these claims are surprising since they narrow the physical domain to what takes place inside the nervous system. This is neither an analytical statement nor something that could be ascertained by philosophical discussion. Rather it is an open empirical issue and should be regarded as such.

Notwithstanding the neuroscientists' faith in a future reduction of mental activity to neural processes, there are still many problems that haven't received any solution. Let me quickly list the biggest obstacles still preventing the mind from being identified with some neural activity:

- Neural activity has only physical properties which are different from phenomenal properties exhibited by our mind — in short David Chalmers' *hard problem* (Chalmers 1996);
- Neural activity *per se* doesn't show any kind of first-person perspective;
- Neural activity is made of many separate firings occurring along relatively long periods of time and in many different locations in the brain. Phenomenal experience shows a pristine unity. This is often called the *binding problem* (Treisman 1998; Revonsuo 1999; Roskies 1999);
- Neural activity has primary and intrinsic properties while our mind has relational and semantic content (Putnam 1975).

Since none of these problems have so far received an intelligible answer, it makes sense to pursue other directions no matter how they could seem counterintuitive to common sense. After all, as Teed Rockwell wrote 'if the brain does not record certain features of a perception that the mind is nevertheless aware of, this must mean that the mind is not identical with the brain' (Rockwell 2005, p. 47). The alternative that we are going to consider in this book and in the remaining part of this chapter is whether the spatio-temporal surrounding of the body could be literally constitutive of our experience and thus of our mind.

3 Externalism by and large

Externalism is a cluster of views holding that the mind is not only the result of what is going on inside the nervous system (or the brain). There are different versions of externalism based either on the kind of relation or on the kind of mental content. Since William Lycan wrote, 'Since Twin Earth was discovered by American philosophical-space explorers in the 1970s' (Lycan 2001, p. 17), externalism subdivided into several variants (various overviews are available in Rowlands 2003; Hurley 2006; Prinz 2009b).

Very roughly, externalism stresses the importance of factors external to the nervous system, but such importance can be articulated and defined in many different ways. There are four important dimensions for evaluating the kind of externalism at hand:

- Phenomenal vs. semantic content;
- Vehicle vs. content scope;
- The extension of the external domain considered;
- Causal vs. constitutive relation.

A few words will clarify each of these dimensions although it is not simply easy to accomplish a clear-cut positioning of each author since often such criteria cut across the work of many authors in a non-linear way.

A first way to classify externalist views is in terms of the kind of mental content. The most common dichotomy is that between phenomenal content and semantic content although other typologies have been considered. For example, enactivist proponents focus on cognitive content or functional states with respect to the agent's behaviour and goals. The cognitive mind overlaps neither with the phenomenal one nor with the collection of intentional/semantic concepts addressed by a given subject. On this issue, David Chalmers has stated clearly that his view of externalism does not take into consideration the phenomenal content, but rather focuses on cognitive aspects only (Chalmers 2008). Thus externalism applies to different kinds of mental content. It is generally held that cognition is easier to deal with than phenomenal experience.

The second classification regards what is the scope of externalism — namely what are we considering, the content or the vehicle of the mind (Hurley 2010)? Consider the concept of a cat. It must rely on some internal machinery and yet it seems to refer to something in the world. The internal machinery is supposedly internal to the body of the agent while the content it refers to is external to it. However, it may not always be the case. It could be argued that the actual process necessary to have a concept requires an active participation with the environment in which that animal lives, doing things with it, interacting with it, and so on. So, the physical underpinnings of it could be located also in the body and even in the environment. Or consider the concept of grasping something. Isn't it embedded into the actual practice? Doesn't it require a first-person experience? So, it is possible to be external only in regard to the content of our mental states (whether they be cognitive, phenomenal, or semantic/intentional) or to require that the vehicles of such mental states extend beyond the limits of the nervous system.

The third classification considers the extension of the mind in functional/physical terms. There are various steps to be considered. At one extreme, some hard-core internalists like Semir Zeki suggest the existence of *microconsciousnesses*, explicit conscious representations implemented using localized dedicated neural circuitry (Zeki 2001b). Each phenomenal experience would thus stem out of a very local neural activity. Similar views have been defended by Koch and other scholars (Quiroga, Kreiman *et al.* 2008). Alternatively, the whole brain or some super-system like the thalamo-cortical system can be held as necessary (Edelman 2001; Tononi 2004). Or it can be argued that the whole nervous system comprehending peripheral nerves is required (Devor 2002). Yet, until the boundary of the skin is not crossed we are still in the scope of internalism. Once the possibility of considering a larger spatio-temporal scope is admitted, the temptation is to go further. The subsequent step is to consider not only the nervous system but also the body as a whole, made of muscles, joints, bones, tendons, various tissues. Is not our cognition embodied in the way in which we move (Varela, Thompson *et al.* 1991/1993; Gallagher and Jeannerod 2002; Gallagher 2005; Pfeifer, Lungarella *et al.* 2007; Menary 2010; Shanahan 2010)? Several authors have considered this possibility somehow rekindled a by certain interpretation of recent findings about mirror neurons (Gallese 2005). And yet the body may not be enough. The tools we routinely use could be considered as part of the subject and thus be integrated into its mental machinery. This is the extended mind hypothesis in a nutshell (Clark and Chalmers 1998; Clark 2008) that will be outlined below. Not only could tools and various devices be integrated, the particular environment in which an agent is acting could be relevant. Cognition could be not only embodied but also situated (Ziemke 2001; Prinz 2005; Gallagher 2009; Robbins and Aydede 2009). Eventually, the right scope could literally contain a large portion of the environment and thus it could swallow a larger temporal span than expected (Manzotti 2006c). It must be remarked that the use of the various terms is not always consistent in the literature and, as a result, they often overlap in vague ways. For instance, embodiment and situatedness are sometimes used as synonyms.

Finally the fourth criterion addresses the kind of dependence between external objects/events/states of affairs and mental states. A rather weak version of externalism suggests that the environment has only a causal and possibly historical role. The external world is necessary only in so far as it causes the brain to grow and develop in a certain way, but once it has reached its working condition (so to speak), the brain is no more in need of having a surrounding world. This is a kind of dependence so weak that it cannot even be considered a legitimate case of externalism. It is the same kind of causal relation that holds between having muscles capable of exerting strength and living in an environment with gravity and inertial masses. Because of the gravity and the need to pull and push objects, the human body develops muscular fibres capable of exerting strength. In the absence of gravity, this process does not take place and muscles quickly lose their strength, as astronauts show. So we can say that it is gravity that causes our muscular strength. But of course, the relation is very weak since, in principle, other kinds of stimuli could have done the same. Furthermore, once the muscles are developed, for a while, they are capable of producing a force without any external intervention — except nutrients. They are sufficient to the occurrence of a force. However, in philosophy of mind the causal relation between the external environment and our mental states is of a stronger kind, even a nomic relation. Other options are some kind of supervenience or some kind of historical/causal legacy not always easily translated into something physically palatable. At the opposite extreme, a very strong kind of externalism claims that the environment is physically constitutive of all kinds of mental experience — phenomenal experience included. Which way does the constitutive relation have to be conceived? This is far from being a self-explanatory notion and all kinds of ontological puzzles are lurking. However, temporarily we can make use of a very simplified physical notion of being constitutive of. So hydrogen and oxygen are both constitutive of water. In this undoubtedly yet-to-be-refined case, x is constitutive of y when x is a necessary part of y (subtle mereological discussion deliberately ignored at this stage). In this simplified scenario, if the environment is constitutive of our mind, it does not mean simply that some kind of supervenience holds between our mental state and the state of affairs in the environment, rather it means that some physical process made by our neural activity plus some other physical events in the environment is identical to what our mental occurrence is.

Merging the various criteria, it is fair to say that the weakest form of externalism is causal content externalism of the semantic version while the most demanding and strongest form is vehicle constitutive externalism extended to phenomenal states also. I will try to outline the various approaches from the point of view of available theories.

3.1 Proto-externalists

To this group belong many authors who weren't considered externalists but whose work suggested views not too far from current forms of externalism. Although this is not the place to outline even briefly their work, it is mandatory to list them.

At the onset of the century there was a brief momentary upsurge of interest in external relations by the group of neorealists (Holt, Marvin *et al.*

1910). In particular, Edwin Holt suggested a view of perception that considered the external world as constitutive of mental content. His rejection of representation paved the way to consider the external object as being somehow directly perceived: 'Nothing can represent a thing but that thing itself' (Holt 1914, p. 142). Holt's words anticipated by almost a century the famous anti-representationalist slogan by Rodney Brooks (Brooks 1991). Recently, neorealist views were refreshed by Francois Tonneau who wrote that 'According to neorealism, consciousness is merely a part, or cross-section, of the environment. Neorealism implies that all conscious experiences, veridical or otherwise exist outside of the brain and are wholly independent of being perceived or not' (Tonneau 2004, p. 97).

To a certain extent, Alfred North Whitehead's process ontology was a form of externalism since it endorsed a neutral ontology, whose basic elements (prehension, actual occasions, events, and processes) seamlessly proceeded from microscopic activity up to the highest level of psychological and emotional life (Whitehead 1929; Griffin 2007).

For one, John Dewey expressed a conception of the mind and its role in the world that is very sympathetic with externalism (Dewey 1925).

More recently, James J. Gibson defended an ecological view of perception and thus of many aspects of the mind. In particular he re-formulated several notions of mental entity which are customarily internal to the brain. Two clear examples are optical flow and information. For Gibson the optical flow is not the computation of the spatial derivatives of the image acquired by the retina as in the classic computational view of the mind championed by David Marr and many others (Marr 1982; Churchland and Sejnowski 1992), rather the optical flow is an environmental dynamic manifold into which the agent is moving. Information gets a twist, too, and it is relocated at an ecological level. Above all, Gibson introduced the notion of affordance, which is external to the agent as such being the potential causal engagement between the body of the agent and some other object (Gibson 1967; Gibson 1979).

Finally, Gregory Bateson must be mentioned, who outlined an ecological view of the mind. Because of his background in cybernetics, he was familiar with the notion of feedback that somehow hampers the traditional separation between the inside and the outside of a system. He questioned the received boundary of the mind and tried to express an ecological view (Bateson 1972/2000; Bateson 1979). Other interesting antecedents of externalism have been reviewed by Shaun Gallagher (Gallagher 2009).

3.2 Semantic externalism

Semantic externalism is the first form of externalism that was dubbed so. As the name suggests it focuses on the mental content of semantic nature (Putnam 1973; Burge 1979). Semantic externalism suggests that mental content does not supervene on what is in the head. Yet the physical basis and mechanisms of the mind remain inside the head. This is a relatively safe move since it does not jeopardize our belief of being located inside our cranium. Hilary Putnam focused particularly on intentionality between our thoughts and external states of affairs—whether concepts or objects (Putnam 1975). In contrast, Tyler Burge emphasized the social nature of the external world, suggesting that semantic content is externally constituted by means of social, cultural, and linguistic interactions (Burge 1979; 1986).

3.3 Phenomenal externalism

Subsequently other authors extended the externalist gist to the phenomenal aspect of the mind. Notably Fred Dretske suggested that 'The experiences themselves are in the head (why else would closing one's eyes or stopping one's ears extinguish them?), but nothing in the head (indeed, at the time one is having the experiences, nothing outside the head) need to have the qualities that distinguish these experiences' (Dretske 1996, p. 144-5). So, although experiences remain in the head, their phenomenal content could depend on something elsewhere. In a similar way, William Lycan defended an externalist and representationalist view of phenomenal experience. In particular, he objected to the tenet that qualia are narrow (Lycan 2001). It has been often held that some, if not all, mental states must have a broad content—that is an external content to their vehicles. For instance, Jackson and Pettit stated that 'The contents of certain intentional states are broad or context-bound. The contents of some beliefs depend on how things are outside the subject' (Jackson and Pettit 1988, p. 381). However, neither Dretske nor Lycan go so far as to claim that the phenomenal mind extends literally and physically beyond the skin. In sum they suggest that phenomenal contents could depend on phenomena external to the body, while their vehicles remain inside.

Another form of content phenomenal externalism is the position dubbed *reflexive monism* by Max Velmans (Velmans 1991; 2000; 2007a,b; 2009). According to Velmans, the object-as-experienced and the experience are the same thing—there is no real difference between phenomenal experience and the physical world we have an experience of. For instance, he suggests that our phenomenal experience of a cat is (roughly) where we perceive the cat to be. He persuasively argues against those authors that maintain that experiences are 'in the brain'. However, he distinguishes 'conscious contents from vehicles which cause or "carry" them' (Velmans 2009, p. 151). Reflexive monism conceives our conscious experience as something which is generated by the information processes in the brain and that is projected onto the physical world. It's a complex view that merges together ontological monism and epistemological dualism (*ibid.*, p. 171). He stresses that our experiences cannot be literally located in the brain, however they are there in so far as they represent the external world which, as far as our experience is concerned, is at the same time psychological and physical. To recap, reflexive monism's key issue seems to be a redefinition of the notion of the physical world in terms of a psychological and phenomenal world. Velmans' trick is to suggest that all the three-dimensional world (all the physical world; *ibid.*, p. 152–4) is part of conscious experience. Experiences and objects are thus coextensive into a phenomenal/physical world. Yet their vehicles remain safely located inside the brain.

3.4 The extended mind

Another sparse group of authors explored another variant of externalism. They observed that the boundaries of cognitive processes are not easily locatable inside the skin. '*Minds are composed of tools for thinking*' (Dennett 2000, p. 21, italics in original). According to Andy Clark, 'cognition leaks out into body and world' (Clark and Chalmers 1998). When someone uses a pencil and paper to compute large sums, cognitive processes extend to the pencil and paper themselves. In a loose sense, nobody would deny it. In a stronger sense, it is rather controversial whether the boundaries of the cognitive mind could extend to the pencil and paper. For most of the proponents of the extended mind, the phenomenal mind remains inside the brain. Commenting on Andy Clark's view, David Chalmers asserts, 'what about the big question: extended consciousness? The dispositional beliefs, cognitive processes, perceptual mechanisms, and moods... extend beyond the borders of consciousness, and it is plausible that it is precisely the nonconscious part of them that is extended' (Chalmers 2008, p. xiv).

3.5 Enactivism and embodiment

Finally, there are many views loosely centred around some form of embodiment that stressed the tight *coupling* between the cognitive processes, the body, and the environment (Varela, Thompson *et al.* 1991/1993; Haugeland 1998; Hurley 1998; Pfeifer 1999; Thelen 2000; Hurley 2003; Pfeifer and Bongard 2006; Prinz 2009b; Robbins and Aydede 2009). Enactivism builds upon the work of other scholars that might be considered as cases of proto-externalism such as Gregory Bateson, James J. Gibson, Merleau-Ponty, Eleanor Rosch and many others. It suggests that the mind is identical to the interactions between the world and the agents.

For instance, Kevin O'Regan and Alva Noë suggested in a seminal paper that the mind is constituted by the sensorimotor contingency between the agent and the world. A sensorimotor contingency is an occasion to act in a certain way and it results from the matching between environmental and bodily properties. To a certain extent sensorimotor contingencies strongly resemble Gibson's affordances. Eventually, Noë developed a more epistemic version of enactivism where the content is the knowledge the agent has as to what it can do in a certain situation. In any case he is an externalist when he claims that 'What perception is, however, is not a process in the brain, but a kind of skilful activity on the part of the animal as a whole. The enactive view challenges neuroscience to devise new ways of understanding the neural basis of perception and consciousness' (Noë 2004, p. 2).

Enactivism received support from various other correlated views such as embodied cognition or situated cognition (Varela, Thompson et al. 1991/1993; Anderson 2003; Chrisley 2003; Gallagher 2005; Pfeifer and Bongard 2006; Chemero 2009; Robbins and Aydede 2009). These views are usually the result of the rejection of the classic computational view of the mind that is centred on the notion of internal representations.

Enactivism receives its share of negative comments particularly from neuroscientists like Christof Koch (Koch 2004, p. 9): 'While proponents of the enactive point of view rightly emphasize that perception usually takes place within the context of action, I have little patience for their neglect of the neural basis of perception. If there is one thing that scientists are reasonably sure of, it is that brain activity is both necessary and sufficient for biological sentience.'

To recap, enactivism is a case of externalism, sometimes restricted to cognitive or semantic aspects, some other times striving to encompass phenomenal aspects. Something that none of the enactivists have so far claimed is that all phenomenal content is the result of the interaction with the environment.

3.6 Recent forms of phenomenal externalism

As it was mentioned at the start, a few authors are now exploring various forms of radical externalism as to phenomenal experience. They have in common the belief that not only cognition but also the conscious mind could be extended in the environment. They also share a common difference with other apparently similar views such as enactivism. While enactivism, at the end of the day, accepts the standard physicalist ontology that conceives the world as made of interacting objects, these more radical externalists consider the possibility that there is some fundamental flaw in our way of conceiving reality and that some ontological revision is indeed unavoidable.

Teed Rockwell recently published a wholehearted attack against all forms of dualism and internalism and proposed that the mind emerges not entirely from brain activity but from an interacting nexus of brain, body, and world. He accused neuroscience of endorsing a form of Cartesian materialism (an indictment issued also in Bennett and Hacker 2003). Dwelling on Dewey's heritage, Rockwell argues that the brain and the body bring to existence the mind as a behavioural field in the environment.

Ted Honderich is perhaps the one with the longest experience in the field. He is defending a position he himself dubbed radical externalism perhaps because of its ontological consequences (Honderich 2004a,b; 2006). One of his main examples is that 'what it actually is for you to be aware of the room you are in, it is for the room a way to exist' (Honderich 2004a). According to him, perceptual consciousness is a way for the world to exist 'Phenomenologically, what it is for you to be perceptually conscious is for a world somehow to exist' (*ibid.*) As a result he identifies consciousness with existence.

One more case of phenomenal externalism is the view called the *spread mind* by Riccardo Manzotti (Manzotti and Tagliasco 2001; Manzotti 2006a). He suggests questioning the separation between subject and object. Such a separation between the world and experience can be set aside because what we consider objects and their phenomenal representations are only two incomplete perspectives and descriptions of the same physical process. This could be done, he argues, by adopting a process ontology that endorses a spread mind physically and spatio-temporally extended beyond the skin. Objects too are no longer autonomous as we know them, but rather actual processes framing our reality (Manzotti 2009).

	Semantic/functional/ cognitive content	Phenomenal content
Content	*Semantic externalism* Putnam 1973; Burge 1979	*Phenomenal externalism* Dretske 1996; Lycan 2001; Tonneau 2004; Byrne and Tye 2006; Tye 2009
Vehicle	*The extended mind* Clark and Chalmers 1998; Dennett 2000; Clark 2008 *Morphological approaches* Pfeifer 1999; Pfeifer and Bongard 2006	
Both content and vehicle	*The embodied mind* Varela, Thompson et al. 1991/1993 *Enactivism* O'Regan 2001; O'Regan and Noë 2001; Hurley 2008; Noë 2009 *Situated cognition* Robbins and Aydede 2009	*Radical externalism* Ted Honderich *Phenomenal vehicle externalism* Manzotti and Tagliasco 2001; Manzotti 2005

Table 1: A taxonomy of different forms of externalism.

4 Conclusion

Externalism is a thought-provoking framework however coarse in its details. Yet, it seems to offer a radical conceptual alternative to the persisting belief in the sufficiency of neural activity for the emergence of phenomenal experience. Any variant of externalism from the less demanding to the most radical ones are going to challenge our current view of what the subject is. Externalism is not a small conceptual step. It challenges the roots of our intuitions as to what we are, the relation between the subject and the object, and the nature of all human experience.

Externalism, of course, is an extremely risky strategy, for it could turn out that it is fundamentally false. However, it is neither an unscientific theory nor an anti-physicalist stance. It is not unscientific since experiments can be devised that could falsify or confirm externalism. It is not anti-physicalist since it does not require anything apart from physical events and processes, albeit different from those admitted by internalists.

It must be noted of course that, at the present state of scientific progress, internalism is by no means less risky being rooted on the opposite hypothesis. The fact that internalism is currently backed up by the confidence of most neuroscientists does not guarantee a better chance of success.

Looking at the theoretical landscape of various forms of externalism, it is interesting to observe that there is a certain tendency to dare to develop more radical versions of it encompassing more and more kinds of mental content and considering stronger kinds of dependence.

In any case the internalism/externalism debate is going to have far reaching effects on our notion of art and of aesthetics. A subject, whose boundaries are not restricted to those of the body, will have a profound impact on what experience and art is. In turn, aesthetics could offer a rich test bed to verify the soundness of many externalist insights as to one of the most intimate and human endeavours—namely art.

References

Adams, D. and K. Aizawa (2008), *The Bounds of Cognition*, Singapore, Blackwell Publishing.
Anderson, M. (2003), 'Embodied Cognition: A Field Guide', *Artificial Intelligence*, **149**: 91–130.
Bateson, G. (1972/2000), *Steps to an Ecology of Mind*, Chicago, The University of Chicago Press.
Bateson, G. (1979), *Mind and Nature: A Necessary Unity*, Cresskill (NJ), Hampton Press.
Bennett, M.R. and P.M.S. Hacker (2003), *Philosophical Foundations of Neuroscience*, Malden (MA), Blackwell.
Block, N.J. (2007), 'Consciousness, Accessibility, and the Mesh Between Psychology and Neuroscience', *Behavioral and Brain Sciences*, **30**: 481–548.
Brooks, R.A. (1991), 'Intelligence Without Representation', *Artificial Intelligence*, **47**: 139–59.
Burge, T. (1979), 'Individualism and the Mental' in P.A. French, T.E. Uehling and H.K. Wettstein, Eds., *Midwest Studies in Philosophy IV*, Minneapolis, University of Minnesota Press: 73–121.
Burge, T. (1986), 'Individualism and Psychology', *Philosophical Review*, **95**: 3–45.

Byrne, A. and M. Tye (2006), 'Qualia Ain't in the Head', *Noûs*, **40** (2): 241–55.
Calvo-Merino, B., C. Jola, *et al*. (2008), 'Towards a Sensorimotor Aesthetics of Performing Art', *Consciousness and Cognition*, **17**: 911–22.
Chalmers, D.J. (1996), *The Conscious Mind: In Search of a Fundamental Theory*, New York, Oxford University Press.
Chalmers, D.J. (2008), 'Foreword to Andy Clark's Supersizing the Mind' in A. Clark, Ed., *Supersizing the Mind*, Oxford, Oxford University Press: 1–33.
Chemero, A. (2009), *Radical Embodied Cognitive Science*, Cambridge (MA), MIT Press.
Chrisley, R. (2003), 'Embodied Artificial Intelligence', *Artificial Intelligence*, **149**: 131–50.
Churchland, P.M. (1985), 'Reduction, Qualia, and the Direct Inspection of Brain States', *The Journal of Philosophy*, **82**: 8–28.
Churchland, P.S. and T.J. Sejnowski (1992), *The Computational Brain*, Cambridge (MA), MIT Press.
Clark, A. (2008), *Supersizing the Mind*, Oxford, Oxford University Press.
Clark, A. and D.J. Chalmers (1998), 'The Extended Mind', *Analysis*, **58** (1): 10–23.
Crick, F. and C. Koch (1998), 'Consciousness and Neuroscience', *Cerebral Cortex*, **8** (2): 92–107.
Currie, G. (2003), 'Aesthetics and Cognitive Science' in J. Levinson, Ed., *The Oxford Handbook on Aesthetics*, Oxford, Oxford University Press: 706–21.
Dehaene, S., J.-P. Changeux, *et al*. (2006), 'Conscious, Preconscious, and Subliminal Processing: A Testable Taxonomy', *Trends in Cognitive Sciences*, **10**: 204–11.
Dennett, D.C. (1995), *Darwin's Dangerous Idea: Evolution and the Meanings of Life*, New York, Simon & Schuster.
Dennett, D.C. (2000), 'Making Tools for Thinking' in D. Sperber, Ed., *Metarepresentations: A Multidisciplinary Perspective*, Oxford, Oxford University Press: 17–29.
Devor, M. (2002), 'Pain Networks' in M.A. Arbib, Ed., *The Handbook of Brain Theory and Neural Networks*, Cambridge (MA), MIT Press: 843–8.
Dewey, J. (1925), *Experience and Nature*, Chicago, Open Court.
Dretske, F. (1995), *Naturalizing the Mind*, Cambridge (MA), MIT Press.
Dretske, F. (1996), 'Phenomenal Externalism, or if Meanings Ain't in the Head, Where are Qualia?', *Philosophical Issues*, **7**.
Eddington, A.S. (1929), *The Nature of the Physical World*, New York, The MacMillan Company.
Edelman, G.M. (2001), 'Consciousness: The Remembered Present', *Annals of the New York Academy of Sciences*, **929**: 111–22.
Faux, S.F. (2002), 'Cognitive Neuroscience from a Behavioral Perspective: A Critique of Chasing Ghosts with Geiger Counters', *Behavior Analyst*, **25**: 161–73.
Finlay, B., R. Darlington, *et al*. (2001), 'Developmental Structure in Brain Evolution', *Behavioral and Brain Sciences*, **24**: 264–308.
Fodor, J.A. (1987), *Psychosemantics: The Problem of Meaning in the Philosophy of Mind*, Cambridge (MA), MIT Press.
Gallagher, S. (2005), *How the Body Shapes the Mind*, Oxford, Oxford Clarendon Press.
Gallagher, S. (2009), 'Philosophical Antecedents of Situated Cognition' in P. Robbins and M. Aydede, Eds., *The Cambridge Handbook of Situated Cognition*, Cambridge, Cambridge University Press.
Gallagher, S. and M. Jeannerod (2002), 'From Action to Interaction', *Journal of Consciousness Studies*, **9** (1): 3–26.
Gallese, V. (2005), 'Embodied Simulation: From Neurons to Phenomenal Experience', *Phenomenology and the Cognitive Sciences*, **4**: 23–48.
Gibson, J.J. (1967), 'New Reasons for Realism', *Synthese*, **17**: 162–72.
Gibson, J.J. (1979), *The Ecological Approach to Visual Perception*, Boston, Houghton Mifflin.
Goldberg, S.C. (2007), *Internalism and Externalism in Semantics and Epistemology*, New York, Oxford University Press.
Goldman, A.I. (1999), 'Internalism Exposed', *The Journal of Philosophy*, **96** (6): 271–93.
Griffin, D.R. (2007), *Whitehead's Radically Different Postmodern Philosophy: An Argument for its Contemporary Relevance*, Albany (NY), State University of New York Press.

Haugeland, J. (1998), 'Mind Embodied and Embedded' in J. Haugeland, Ed., *Having Thought: Essays in the Metaphysics of Mind*, Cambridge (MA), Harvard University Press.
Heisenberg, W. (1958), *Physics and Philosophy*, New York, Harper & Row.
Hirose, N. (2002), 'An Ecological Approach to Embodiment and Cognition', *Cognitive Systems Research*, **3**: 289–99.
Holt, E.B. (1914), *The Concept of Consciousness*, New York, MacMillan.
Holt, E.B., W.T. Marvin, et al. (1910), 'The Program and First Platform of Six Realists', *The Journal of Philosophy, Psychology and Scientific Methods*, **7**: 393–401.
Honderich, T. (2004a), 'Consciousness as Existence, Devout Physicalism, Spiritualism', *Mind and Matter*, **2** (1): 85–104.
Honderich, T. (2004b), *On Consciousness*, Edinburgh, Edinburgh University Press.
Honderich, T. (2006), 'Radical Externalism', *Journal of Consciousness Studies*, **13** (7–8): 3–13.
Hurley, S.L. (1998), *Consciousness in Action*, Cambridge (MA), Harvard University Press.
Hurley, S.L. (2003), 'Action, the Unity of Consciousness, and Vehicle Externalism' in A. Cleeremans, Ed., *The Unity of Consciousness: Binding, Integration, and Dissociation*, Oxford, Oxford University Press.
Hurley, S.L. (2006), 'Varieties of Externalism' in R. Menary, Ed., *The Extended Mind*, Aldershot, Ashgate Publishing.
Hurley, S.L. (2008), 'The Shared Circuits Model: How Control, Mirroring and Simulation Can Enable Imitation, Deliberation, and Mindreading', *Behavioral and Brain Sciences*, **31** (1): 1–22.
Hurley, S. (2010), 'The Varieties of Externalism' in R. Menary, Ed., *The Extended Mind*, Cambridge (MA), MIT Press: 101–55.
Iseminger, G. (2005), 'Aesthetic Experience' in J. Levinson, Ed., *The Oxford Handbook of Aesthetics*, Oxford, Oxford University Press: 99–116.
Jackson, F. and P. Pettit (1988), 'Functionalism and Broad Content', *Mind*, **97** (387): 381–400.
Kim, J. (1995), 'Mental Causation: What? Me worry?', *Philosophical Issues*, **6**: 123–51.
Kim, J. (1998), *Mind in a Physical World*, Cambridge (MA), MIT Press.
Koch, C. (2004), *The Quest for Consciousness: A Neurobiological Approach*, Englewood (CO), Roberts & Company Publishers.
Koch, C. (2010), 'An Answer to the Riddle of Consciousness', *Scientific American*, **303** (3): 76.
Lamme, V.A.F. (2003), 'Why Visual Attention and Awareness are Different', *Trends in Cognitive Sciences*, **7** (1): 12–9.
Lamme, V.A.F. (2006), 'Towards a True Neural Stance on Consciousness', *Trends in Cognitive Sciences*, **10** (11): 494–501.
Levinson, J. (2005), 'Philosophical Aesthetics: An Overview', *The Oxford Handbook of Aesthetics*, Oxford, Oxford University Press: 3–24.
Llinàs, R. (2001), *I of the Vortex: From Neurons to Self*, Cambridge (MA), MIT Press.
Lycan, W.G. (2001), 'The Case for Phenomenal Externalism' in J.E. Tomberlin, Ed., *Philosophical Perspectives, Vol. 15: Metaphysics*, Atascadero, Ridgeview Publishing: 17–36.
Manzotti, R. (2005), 'Outline of an Alternative View of Conscious Perception', *Towards a Science of Consciousness 2005 (TSC2005)*, Copenhagen.
Manzotti, R. (2006a), 'An Alternative Process View of Conscious Perception', *Journal of Consciousness Studies*, **13** (6): 45–79.
Manzotti, R. (2006b), 'Consciousness and Existence as a Process', *Mind and Matter*, **4** (1): 7–43.
Manzotti, R. (2006c), 'A Process Oriented View of Conscious Perception', *Journal of Consciousness Studies*, **13** (6): 7–41.
Manzotti, R. (2006d), 'A Radical Externalist Approach to Consciousness: The Enlarged Mind' in A. Batthyany and A. Elitzur, Eds., *Mind and Its Place in the World: Non-Reductionist Approaches to the Ontology of Consciousness*, Frankfurt, Ontos-Verlag: 197–224.
Manzotti, R. (2009), 'No Time, No Wholes: A Temporal and Causal-Oriented Approach to the Ontology of Wholes', *Axiomathes*, **19**: 193–214.
Manzotti, R. and P. Moderato (forthcoming), 'Is Neuroscience the Forthcoming "Mindscience"?', *Behaviour and Philosophy*.

Manzotti, R. and V. Tagliasco (2001), *Coscienza e Realtà: Una Teoria della Coscienza per Costruttori e Studiosi di Menti e Cervelli*, Bologna, Il Mulino.
Marr, D. (1982), *Vision*, San Francisco, Freeman.
Melzack, R. (1993), 'Pain: Past, Present, and Future', *Canadian Journal of Experimental Psychology*, **47** (4): 615-29.
Menary, R., Ed. (2010), *The Extended Mind*, Cambridge (MA), MIT Press.
Mendola, J. (2008), *Anti-Externalism*, Oxford, Oxford University Press.
Noë, A. (2004), *Action in Perception*, Cambridge (MA), MIT Press.
Noë, A. (2009), *Out of the Head: Why You Are Not Your Brain*, Cambridge (MA), MIT Press.
Noë, A. and E. Thompson (2004), 'Are There Neural Correlates of Consciousness?', *Journal of Consciousnesss Studies*, **11** (1): 3-28.
O'Regan, K. (2001), 'What it is Like to See: A Sensorimotor Theory of Perceptual Experience', *Synthese*, **129**: 79-103.
O'Regan, K. and A. Nöe (2001), 'A Sensorimotor Account of Visual Perception and Consciousness', *Behavioral and Brain Sciences*, **24** (5): 939-1011.
Pfeifer, R. (1999), *Understanding Intelligence*, Cambridge (MA), MIT Press.
Pfeifer, R. and J. Bongard (2006), *How the Body Shapes the Way We Think: A New View of Intelligence*, New York, Bradford Books.
Pfeifer, R., M. Lungarella, et al. (2007), 'Self-Organization, Embodiment, and Biologically Inspired Robotics', *Science*, **5853** (318): 1088-93.
Prinz, J. (2005), 'Is Consciousness Embodied?' in P. Robbins and M. Aydede, Eds., *The Cambridge Handbook of Situated Cognition*, Cambridge, Cambridge University Press: 1-20.
Prinz, J. (2009a), *The Conscious Brain*, New York, Oxford University Press.
Prinz, J. (2009b), 'Is Consciousness Embodied?' in P. Robbins and M. Aydede, Eds., *The Cambridge Handbook of Situated Cognition*, Cambridge, Cambridge University Press: 419-36.
Putnam, H. (1973), 'Meaning and Reference', *The Journal of Philosophy*, **70** (19): 699-711.
Putnam, H. (1975), *Mind, Language, and Reality*, Cambridge, Cambridge University Press.
Pylkannen, P. (2007), *Mind, Matter and the Implicate Order*, Berlin, Springer.
Quiroga, R.Q., G. Kreiman, et al. (2008), 'Sparse but not "Grandmother-cell" Coding in the Medial Temporal Lobe', *Trends in Cognitive Sciences*, **12**: 87-92.
Rakic, P. (2009), 'Evolution of the Neocortex: A Perspective from Developmental Biology', *Nature Reviews Neuroscience*, **10** (10): 724-36.
Ramachandran, V.S. and W. Hirstein (1999), 'The Science of Art: A Neurological Theory of Aesthetic Experience', *Journal of Consciousness Studies*, **6**: 15-51.
Revonsuo, A. (1999), 'Binding and the Phenomenal Unity of Consciousness', *Consciousness and Cognition*, **8**: 173-85.
Revonsuo, A. (2010), *Consciousness: The Science of Subjectivity*, Hove, Psychology Press.
Robbins, P. and M. Aydede, Eds. (2009), *The Cambridge Handbook of Situated Cognition*, Cambridge, Cambridge University Press.
Rockwell, T. (2005), *Neither Ghost nor Brain*, Cambridge (MA), MIT Press.
Roskies, A.L. (1999), 'The Binding Problem', *Neuron*, **24**: 7-9.
Rowlands, M. (2003), *Externalism. Putting Mind and World Back Together Again*, Chesham, Acumen Publishing Limited.
Seth, A.K., E. Izhikevich, et al. (2006), 'Theories and Measures of Consciousness: An Extended Framework', *Proceedings of the National Academy of Sciences of the United States of America*, **103**: 10799-804.
Shanahan, M.P. (2005), 'Global Access, Embodiment, and the Conscious Subject', *Journal of Consciousness Studies*, **12** (12): 46-66.
Shanahan, M. (2010), *Embodiment and the Inner Life: Cognition and Consciousness in the Space of Possible Minds*, Oxford, Oxford University Press.
Strawson, G. (2005), 'Why Physicalism Entails Panpsychism' in *Proceedings of TSC2005*, Danish National Research Foundation.
Thelen, E. (2000), 'Grounded in the World: Developmental Origins of the Embodied Mind', *Infancy*, **1** (1): 3-28.

Thompson, E. and F.J. Varela (2001), 'Radical Embodiment: Neural Dynamics and Consciousness', *Trends in Cognitive Sciences*, **5** (10): 418–25.
Tonneau, F. (2004), 'Consciousness Outside the Head', *Behavior and Philosophy*, **32**: 97–123.
Tononi, G. (2004), 'An Information Integration Theory of Consciousness', *BMC Neuroscience*, **5** (42): 1–22.
Tononi, G. (2008), 'Consciousness as Integrated Information: A Provisional Manifesto', *Biological Bulletin*, **215**: 216–42.
Tononi, G. and C. Koch (2008), 'The Neural Correlates of Consciousness: An Update', *Annals of the New York Academy of Sciences*, **1124**: 239–61.
Treisman, A. (1998), 'The Binding Problem', *Current Opinion in Neurobiology*, **6**: 171–8.
Tye, M. (2009), 'Phenomenal Externalism' in M. Tye, Ed., *Consciousness Revisited*, Cambridge (MA), MIT Press: 193–200.
Varela, F.J., E. Thompson, *et al.* (1991/1993), *The Embodied Mind: Cognitive Science and Human Experience*, Cambridge (MA), MIT Press.
Velmans, M. (1991), 'Is Human Information Processing Conscious?', *Behavioral and Brain Sciences*, **14**: 651–69.
Velmans, M. (2000), *Understanding Consciousness*, London, Routledge.
Velmans, M. (2007a), 'Reflexive Monism', *Journal of Consciousness Studies*, **15** (2): 5–50.
Velmans, M. (2007b), 'Where Experiences Are: Dualist, Physicalist, Enactive and Reflexive Accounts of Phenomenal Consciousness', *Phenomenology and the Cognitive Sciences*, **6**: 547–63.
Velmans, M. (2009), *Understanding Consciousness (Second Edition)*, London, Routledge.
Whitehead, A.N. (1925), *Science and the Modern World*, New York, Free Press.
Whitehead, A.N. (1929), *Process and Reality*, London, Free Press.
Zeki, S. (2000), *Inner Vision: An Exploration of Art and Brain*, Oxford, Oxford University Press.
Zeki, S. (2001a), 'Artistic Creativity and the Brain', *Science*, **293**: 51–2.
Zeki, S. (2001b), 'Localization and Globalization in Conscious Vision', *Annual Review of Neuroscience*, **24**: 57–86.
Zeki, S. (2002), 'Neural Concept Formation & Art', *Journal of Consciousness Studies*, **9**: 53–76.
Zeki, S. (2003), 'The Disunity of Consciousness', *Trends in Cognitive Sciences*, **7** (5): 214–8.
Zeki, S. (2009), *Splendours and Miseries of the Brain: Love, Creativity and the Quest for Human Happiness*, London, Wiley.
Ziemke, T. (2001), 'The Construction of Reality in the Robot Constructivist Perspective on Situated Artificial Intelligence and Adaptive Robotics', *Foundations of Science*, **6**: 163–233.

Erik Myin & Johan Veldeman

Externalism, Mind, and Art

According to externalists about the mind, the mind is not confined to the head. Instead the mind is said 'to spread out in' or 'to be partially constituted by' the world. In this chapter, we distinguish the different ways in which these claims have been elaborated, differentiating between *active externalism* and *explanatory externalism*. Then we turn to the question to which extent art production and perception can be illuminated by these externalisms. Finally, we consider the debate between internalists and contextualists in analytical aesthetics, and we argue that externalism about the mind contains the necessary insights to dissolve the opposition between perceptual experience and context which lies at the heart of that debate.

1 Externalism about the mind

Consider climbing a ladder. While sliding your hand over the sides of the ladder, you place your feet on the rungs, only to move one foot to a higher rung. You take great care not to let go of the ladder, or to lose contact completely. If the ladder would fall, you know you would be in trouble. You and the ladder are tightly coupled, and you take care to maintain that tight level of interaction. Undoubtedly, ladder climbing involves a complicated causal interaction between two separable entities, the climber and the ladder. Moreover, one can, abstracting from the actual empirical process, consider each entity's contribution apart from the other entity's contribution: the support the rungs provide, and the muscular processes in the body. Yet it would be far fetched to adopt an *internalist* stance towards the process of ladder climbing, according to which all that really matters to that process is internal to the body. For the process of ladder climbing seems, in order to be what it is, to require, besides the muscular processes, the ladder, just as the process of drinking milk requires milk, besides all the bodily movements of drinking.

But are mental processes—perceiving, thinking, judging, aesthetically appreciating, to mention only a few—comparable to ladder climbing? Do these involve the world in that essential sense in which ladder climbing involves the world? *Externalists* about the mind think so, opposing themselves against those in the internalist tradition in philosophical and scientific thinking about the mind. According to that tradition, the mind is special precisely because it is inner. Mental processes of course involve causal interactions with the world, but the genuinely *cognitive* or *mental* aspects of the interaction happen within the boundaries of the body — indeed (almost invariably) within the brain.

This is not the place to provide a complete review of the internalism/externalism debate in the philosophy of mind. Instead, we will present a type of case in which we think externalism about the mind is particularly strong — a case in which mental processes seem genuinely like the process of climbing a ladder. We will show how two different kinds of externalism can be developed from such a paradigmatic case, in such a way that it remains defensible against the standard internalist critiques. Then, we will consider the extent to which either applies to art and its appreciation. This discussion will start with art production. As art production cries out for an analysis along externalist lines, and as art production is central to art appreciation — aesthetics — the latter too should be seen as essentially world-involving. We will then show how the adoption of an externalist stance can help to resolve a tension in the *prima facie* somewhat different internalism/externalism debate in analytical aesthetics — a tension which arises because of an unnecessary opposing of experience and context. Externalism about the mind, so we will argue, allows us to dissolve this tension.

1.1 Active externalism: An example

Consider the following case. There is a child that is not yet able to do sums in her head. She can read the number characters, however, and she can group cubes in groups of two, three or four. Imagine she is sitting on a table with cube-shaped blocks. Written on the wall there is a line, on which numbers are marked. The distance between two consecutive numbers is exactly the height of a block. Now this child solves questions about sums such as 'How much is two plus three?' simply by collecting one group of two, and one of three blocks, by then putting them on top of each other near the vertical line, and by simply reading off the number written there where her construction reaches. In this case, it seems quite hard to deny that this child has computed the sum by manipulating objects in the world. Moreover, it is hard to deny that the process of adding two to three *is* to the physical manipulation of the blocks.

Interestingly, she has not done something with the blocks that she would be able to do without them — as she cannot yet do sums 'in the head'. Take away the blocks, or the number line for that matter, and the process of adding loses its foothold. Therefore, whatever it is that inner processes might have contributed to the process of solving the sum by manipulating the blocks, this couldn't have been the process of computing the sum *itself*. The internalist thesis that whatever is genuinely cognitive about the process remains confined to the head runs against the basics of this story.

This type of case, we think, is what almost any externalist about the mind would regard as a paradigmatic 'extended' cognitive process, in the sense that it is a primary example of a kind of process which is partly 'constituted' by activities involving worldly elements. The cognitive process is what it is because of, and cannot be described without mentioning, the actions out there on non-bodily entities: the manipulation of the blocks. Moreover, the interaction with the blocks is not merely a causal prelude to genuine cognitive processes inside, because — as mentioned just above — there is no internal process of computing the sum.

1.2 Parity and beyond

We have made up the example of the child that sums by piling up blocks especially for the current occasion. The externalism about the mental which it illustrates found its way into the debate mainly through two other notorious examples. These were introduced in one of the most discussed papers in recent philosophy, 'The Extended Mind', written by Andy Clark and David Chalmers (Clark and Chalmers 1998). In it, Clark and Chalmers distinguished 'active externalism', according to which the mind is 'constituted partly by features of the environment', from 'traditional externalism' *à la* Putnam or Burge, according to which the meaning of words such as natural kind terms are determined by the natural or social environment — 'water' meaning something different when the waterlike substance in the environment has a different composition than H_2O.[1]

Though, as we will see in a moment, Clark and Chalmers' paper contains many strands for externalism about the mind, there is, in their origi-

---

[1] Traditional externalism was derived from considerations about the meaning of a certain class of referring terms. Thus according to Putnam's so-called 'Twin-Earth' thought experiment, we are proposed to imagine a planet that is just like earth except in one respect: H_2O is replaced by a chemical substance with the molecular structure XYZ, which has all the observable characteristics of water. So whereas on earth lakes and rivers are filled with H_2O, on Twin Earth they are filled with XYZ. English-speaking Twin-Earthians use their expression 'water' in exactly the way we use 'water'. But our 'water' and Twin-Earthian 'water' have different extensions: the first refers to H_2O, whereas the second refers to XYZ. So, when my Twin-Earth counterpart and I both use the expression 'water', our expressions differ in meaning, even if we happen to be in exactly the same brain state and if our speech behaviour happens to be indistinguishable. And, so goes the argument, if linguistic meaning is necessarily bound up with one's causal relations to certain natural substances, then the same holds for the meanings of thoughts. Active externalism is importantly different: it asserts that the environment can play an active role in constituting and driving cognitive processes. It is a claim not merely about the determination of mental content, but also about the location of mental processes. According to active externalism, these extend beyond the skin of the individual.

nal paper, a clearly discernable focus on defending active externalism by relying on the *parity principle*. This gets formulated as:

> If, as we confront some task, a part of the world functions as a process which, *were it done in the head*, we would have no hesitation in recognizing as part of the cognitive process, then that part of the world *is* (so we claim) part of the cognitive process. (Clark and Chalmers 1998, p. 8)

Clark and Chalmers have given two well-known examples to which it was meant to apply, in favour of an active externalism about cognitive processes. The first one concerned the computer game Tetris, in which falling shapes can be rotated so as to fit into already standing shapes, all seen on a computer screen. People playing the game have to judge which falling shapes will fit, after possible rotation. Such a judgment can be based on mental imagery, an example of an internal process if there ever was one. But Clark and Chalmers remarked that instead of 'mentally rotating an image', players could be manipulating a physical image on a computer screen to the same effect. The physical image would perform the same cognitive function as the internal image. A second, particularly iconic example featured Otto, a 'slightly amnesic' subject. Instead of being able to rely as fully on his natural memory as a non-amnesic subject would, Otto relies on a notebook, in which he writes down facts he anticipates will be important to him in the future, but which he knows he is likely to forget. Whenever the need arises, for example, when he needs to find his way to the Museum of Modern Art in New York, Otto accesses the note in his booklet, rather than accessing a memory trace in his brain.

If the notebook is constantly accessible to Otto, Clark and Chalmers defend, the inscriptions in the notebook serve as well as the 'vehicle' of memory as some neural memory trace. There is parity between those two, and it would be chauvinistic to consider the neural 'vehicle', but not the note as constitutive of Otto's remembering.

But parity, and with it the pair of established examples for active externalism, should be treated with great caution. Other philosophers with a broadly 'active externalist' mindset have, with hindsight, remarked that relying on the parity principle as the main motivation for externalism about the mind is quite problematic (Menary 2006, 2007, 2010; Sutton 2010). The most important reason for this is that the parity principle seems to restrict cases of (partially) external mental processes to those in which the externalized process is strongly similar to an internal process. It seems odd to impose this restriction, as the most interesting world-involving processes seem to be *augmentative*: processes in which the reliance on the external, or the actions or manipulations in the non-bodily environment, make possible something which wasn't possible without them. That is, if the external means are merely used to perform a task that could be done just as well internally, they add little. The external manipulation amounts to nothing more than a different way of doing the same as what was previously done internally. This remains true if the external world, as is often claimed, is seen as 'reducing memory load' or 'increasing computational resources' (Clark 2003). In such a picture the external world figures in a purely quantitative way, supplying more of what is only to a limited extent available internally. It is exactly when the externally enabled processes are *not* similar, that they become genuinely valuable. Then they broaden the spectrum of cognitive processes. Note that this augmentative aspect is at the heart of our summing child example.

There seems to be a convergence notable in the writings of different theorists defending externalism about the mind towards this kind of augmentative approach. Called 'integrationism' by Menary (2007; 2010), or 'the second wave of extended mind research' by Sutton (2010), it rejects parity as a basis and stresses the novelty brought about by processes constitutively relying on the world. This is not to say that it is necessarily in conflict with all that was initially proposed by Clark and Chalmers when they launched 'active externalism' in Clark and Chalmers (1998). In retrospect, that paper contained the seeds of both parity-based and augmentative variants of externalism about the mind. For example, the way Clark and Chalmers looked at language was certainly very much along the lines of the augmentative consensus. For they wrote:

> Without language, we might be much more akin to discrete Cartesian 'inner' minds, in which high-level cognition relies largely on internal resources. But the advent of language has allowed us to spread this burden into the world. Language, thus construed, is not a mirror of our inner states but a complement to them. It serves as a tool whose role is to extend cognition in ways that on-board devices cannot. Indeed, it may be that the intellectual explosion in recent evolutionary time is due as much to this linguistically-enabled extension of cognition as to any independent development in our inner cognitive resources. (Clark and Chalmers 1998, p. 18)

In his later solo works, Clark has heavily stressed the 'transformative potential' of external 'tools', which range from spoken and written words and other symbols, to by now ordinary things such as laptops, and actual or possible bodily extensions such as brain implants (Clark 2003). One example presented and discussed by Clark which is particularly interesting in this context concerns the production of abstract artworks. It refers to the work done by van Leeuwen, Verstijnen and Hekkert (1999) on the role of sketchpads in the production of certain forms of abstract art. Van Leeuwen and his colleagues showed that the creation of multiply interpretable elements in drawing, typical in much of abstract art, is essentially dependent on the use of external sketches. They claimed that this is due to the difficulty of holding several and simultaneous interpretations of images in the mind, which was demonstrated by Chambers and Reisberg

(1985). The latter have shown that people cannot flip to the other of the duck/rabbit pair,[2] if their only means to hold the image is mental. When looking with the mind's eye, either the rabbit or the duck are seen, without the possibility of switching. It is only by looking at an external image that its many possible interpretations can be generated and combined in a reliable enough way to guide a design process. Thus, the use of a sketchpad opens up possibilities for new forms of artistic creation which would not have existed without them (Clark 2003). What emerges then is a form of extended cognition in which established elements and operations are closely integrated with additional ones in order to form a single novel 'unit of cognitive analysis' (Menary 2006).

It should be emphasized that this broader, non-parity based, externalism about the mind is — unlike its parity based kin — able to withstand the charge by internalist opponents that it commits a 'coupling constitution fallacy', by mistaking a causal contribution of the world for a constitutive one (Adams and Aizawa 2001; 2008). That is, internalists like Adams and Aizawa of course do not deny that cognitive and perceptual processes often involve causal interaction with the world. Yet they claim that all that is properly *cognitive* in such interactions happens internally. The mind receives input from the world, and sends outputs back to the environment, but it is in the internal 'processing' of such input, and the preparation of the output, that the genuinely cognitive episodes happen. Such a line of reasoning might appear plausible in a parity case, as this starts from the assumption that the externalized processes merely mimic already existing internal processes. Internalists will insist that *only* those internal processes are genuinely cognitive, while the broader external process is only derivatively cognitive. While this reasoning is not unproblematic in itself, because it takes as given a cognitive/non-cognitive boundary that coincides with an internal/external boundary (Hurley 2010), clearly its application to a non-parity based externalism seems blocked, because there might not be *any* internal component that has the required cognitive properties in its own right. That is, the cognitive process exists only in its spread out, world-involving form, not in an internal plus an external variety. Remember how this is true for our initial example of the child that sums by piling up blocks. Apart from the external world involving process, there is no process of addition going on, let alone an internal one. But it is true of Clark's example of abstract art too: apart from the process involving sketchbooks, there is no internal cognitive process devising multi-interpretable images. In fact, given the empirically established constraints on internalistic imagery, there *could not even be* an internalist process of that kind.

2 The famous duck/rabbit image shows a figure that can both be interpreted as a duck or as a rabbit, though not simultaneously as both. That is, if one stares at the image, one's experience continually changes from the one to the other interpretation.

So far our examples for active externalism, integration or transformation all involve external entities, or artefacts: the image and the note in the booklet in the parity cases, the blocks and the sketchpad in the other cases. But, as again remarked by a number of theorists of the externalist persuasion, nothing obliges to restrict active externalism to situations in which artefacts, or concrete objects, are involved (Hurley 2010; Wilson 2010). The parity principle, however—in its original formulation cited above—might dispose towards assuming the involvement of artefacts or objects. It directs toward objects because the parity principle is based on the idea of a similar functional role played by *something*. *Something external* is supposed to stand in for—play the functional role of—*something* internal. At least on the face of it, it therefore seems to commit to an object-based model of the cognitive capacities at issue. This gets confirmed by the iconic examples given in the original paper by Clark and Chalmers, which involve respectively images on a computer screen (standing in for 'mental images') and inscriptions in a notebook replacing 'memory traces'. And the objects at play tend to be artefacts, again because of parity: something *novel* external has to replace what normally is internal. Lines in a notebook replace memory traces or images on a screen replace visual imagery.

Therefore, once parity is abandoned, there seems to be no reason to require the presence of artefacts or even objects in order to make an external world involving process cognitive or mental. In particular, *perception* can rely on the world in an essential, constitutive manner, without any mediating man-made artefacts.

At least some cases of perception might come about because an animal is sensitive to the patterns of change it creates through self-movement (Chemero 2010 offers insightful elaborations of this theme, originated by Gibson 1966). To stick with a well studied subject, the changing distance between oneself and an object or a surface one approaches, and therefore the time before one will collide with it, might be directly read off from the spatial and temporal way in which the retinal image expands. As you approach a brick wall, for example, there will be specific patterns of change that occur in the retinal image over time. The projection of the boundaries of each separate brick will expand and 'flow outwards', and smaller scale texture will become visible. Traditional internalist models for this phenomenon invoke internal computations carried out over subsequent representations of what is sampled at the receptor level and sent inwards for further processing. An externalist can claim that at least part of the computational work is not performed by internal neural mechanisms on representations of the stimuli, but directly on the self-generated changing pattern in the sampled light (Wilson 2004; 2010). The animal literally sees by acting.

Such crucial patterns of changes in stimulus upon self movement have been called 'sensorimotor contingencies' in an approach to perception which has proven to be very appealing to externalists, namely the 'sensorimotor contingency theory' (O'Regan and Noë 2001; Myin and O'Regan 2009; Hurley 2010; Wilson 2004; 2010). According to this theory, vision should be seen as a way of doing something in an environment, as a way of interaction, rather than as a process that leads to some end product — some 'percept' generated in and by the brain. Seeing a line, on a sensorimotor account, should not be thought of as the construction of a line-like or line-isomorphic representation, but can be understood as actively finding out that the retinal stimulus doesn't change if one moves one's eyes along the line.

Because, certainly from a very young age on, seeing is possible without any movement, the sensorimotor approach needs to be expanded to account for vision in general. On the other hand, motion might be more important in more circumstances than might initially be thought (Findlay and Gilchrist 2003). For example, there are strong indications that the seeing of detail heavily relies on world-involving interactive processes. It is an established fact that, when looking at a scene in a natural context, such as making coffee in a kitchen, we only look at those objects in the scene which are important to our current needs (Land 2004; see Myin and O'Regan 2009 for more on this theme, which is very much manifest in reading as well). Instead of having the whole scene simultaneously before our mind, we remain in touch with the whole of it, by connecting, over time, with those elements that matter to us at that particular moment. Such a process, also present in reading, has been compared to holding a bottle: strictly speaking one's tactile contact is limited to where the fingers touch the bottle, yet you feel the whole bottle. Similarly, while strictly speaking we are at any time only looking at a very limited part of the scene, we nevertheless have the impression of seeing the whole scene. How come? According to the sensorimotor approach, a key factor in explaining this involves the fact that our visual perceptual system is such that it is immediately grabbed by significant changes in the scene. If some important visual change appears in a part of the scene before you to which you are not currently focusing, you will be attracted to that change, and thereby you'll move your eyes, head or body to focus on it. Therefore, under normal circumstances, no significant change will escape your attention. You will hold the scene in a visual grip.

But you do so by being active, by literally turning to what is probably of interest. You and the scene become knit together in your activity of perceiving it, just as the ladder and the climber become tightly coupled in the process of ladder climbing.

1.3 Active externalism: Summing up

Relying on a number of examples, we have given substance to the core idea of active externalism about the mind, namely that the environment plays a constitutive role for some cognitive or mental processes. We have indicated the benefits from not basing active externalism on parity. One of those is that it then applies, beyond more high-level cognitive processes such as calculating, to perception.

1.4 Explanatory externalism

The active externalism we have just sketched was developed with online cognitive or, by extension, perceptual activity in mind. The environment or body-external objects or tools play an essential role in ongoing cognitive processes. The role is essential because no internal process plays that role (the denial of parity), and therefore the process would not take place, or be possible, without the external environment. In the way we have sketched active externalism, we have taken this sense of playing an essential role to be what 'constitutive' means.

But it is not necessary to restrict externalism to online processes. A more encompassing externalism is possible, according to which not only the current but also the past involvement of external elements bestows extended status to cognitive or mental phenomena. Susan Hurley, in her 'The Varieties of Externalism', defends precisely such 'explanatory externalism' (Hurley 2010; see also Hurley 1998). Broadly, the core idea here is that a current cognitive or mental phenomenon is partly constituted by environmental factors, if these environmental or external (rather than internal) factors, through their role in the past or the present, are necessary to explain why the cognitive or mental phenomenon is what it is. One example she gives, building on a joint paper with Alva Noë (Hurley and Noë 2003), concerns neural plasticity: the phenomenon whereby a certain part of the brain acquires, under changed circumstances, a different role. The part of the cortex that is visual in seeing people, for example, seems to be involved in tactile experience in blind people (see Hurley and Noë 2003 for many other examples). It seems difficult, if not straightforward impossible, to explain this fact by merely invoking brain-internal, strictly neural properties. Rather, the explanation that this case cries out for is an account which essentially involves the world, in an historical/developmental way. It is because the 'visual cortex' of blind people has been recruited in their — much more comprehensive (think about Braille reading) — tactile interactive dealings with the world, that the 'visual cortex' of the blind has become like the tactile, and unlike the visual cortex of the seeing.

Explanatory externalism clearly significantly broadens the range of application for externalism about the mind. Some have expressed the concern that it might thereby become trivial, because of the breadth of its scope (for example Rowlands 2009). However, this criticism seems to conflate ubiquity, which concerns reality, with triviality, which is epistemic. That is, even if explanatory externalism applies broadly to reality (*if* it does), it still might not be trivial because it is widely believed not to apply so broadly, or because it is believed in a number of cases not to apply, where it in fact does apply. Many theorists are explanatory internalists about all or at least some aspects of the mind. They might have principled and very general reasons for that, or have specific internalist explanations for concrete phenomena in mind. Fodor's methodological solipsism, according to which all explanations in psychology sidestep reality, because only reality as mentally represented counts for explanatory purposes in psychology, is an example of the former (Fodor 1980). As an example of the latter, take the common claim that perhaps the phenomenal character of colour is to be explained by specifics of neural coding. That red feels like red, for example that it feels opposed to green, would then be explained, not by any internal factor relating to the external stimulus, but purely by the specific ways in which the human colour system is wired up—namely along the lines of 'opponent channels', among an opponent green/red channel which signal either green or red, but never both at once (Hardin 1988). The existence of competing—and quite widespread—contrasting positions shows explanatory externalism is far from trivial.

1.5 Externalism (further) challenged and defended

Critics of externalism, of which there are plenty, often hammer on the fact that dependency on external elements is not sufficient to prove constitutive involvement of the external in the mental (Adams and Aizawa 2001; 2008; Rupert 2004; 2010). They think such a criticism applies both to active as well as to explanatory externalism, as it applies both to dependency relations in the here and now as in the past. The fact that one uses an external element in one's cognitive process doesn't make, so internalists claim, that external element a literal part of the cognitive process. *A fortiori* an external element doesn't become part of a cognitive process if it has played some past role in the shaping of how that cognitive process now functions.

This criticism can seem to make most sense if the meaning of 'constitutive' driving it is one of location. It might indeed seem that the (especially past) dependence on external aspects does not make those aspects part of a current cognitive process. As one learns one's language from other people, by that line of reasoning, all the people that have played a part in learning a language would become part of every exercise of one's language faculty (Rupert 2010).

But this internalist critique restricts externalism to a thesis about localization. This clearly is not necessary, as is strongly argued for by Hurley

(2010). Hurley favours a sense of 'constitutively involved' which makes explanatory role primordial. Something A, then, is constitutively involved in B if A explains B. But one can opt as well for a metaphysical reading of constitutive — as was done at times by us in the above — by reading 'constitutive' as 'essential' — as 'what is needed to make something into what it is'. Moreover, both readings could converge, as metaphysical issues about what is essential for X might be decided by facts about what explains X (Ross and Ladyman 2010; Hurley 2010).

This is not the place to resolve these issues (if they can be resolved at all — see Chemero 2010). In any case, they need not be resolved in order to go on with our enterprise with the sufficiently definite meaning(s) we have now for externalism about the mind.

So we can go on to ask: in what sense is art externalist — in the senses discussed?

2 From mind to art and its appreciation

Our consideration of art and its appreciation from an externalist perspective will proceed in two steps. First, we will indicate some important ways in which both art production and art consumption seem to call for an account along the lines of either active or explanatory externalism. Second, we will connect to a *prima facie* somewhat different internalist/externalist debate in analytical aesthetics. We will show how the externalist position regarding the mind can lead to a balanced resolution of this debate.

2.1 Externalist aspects of art production and art consumption

The example about the role of sketchpads in the production of abstract art highlights how the process of actually *producing* the art is an externalistic activity. Worldy elements are directly and essentially involved in the process of production, without them having any internal counterparts. Indeed, because of psychological or neurobiological limitations, the process *couldn't* be done internally (Chambers and Reisberg 1985). Through the world-involving loops something novel is established which would not be possible without them. All the conditions for the world to be constitutively involved in the process of making abstract art are fulfilled. This example thus shows a case which unfolds along active externalist lines.

To what extent can the example be generalized? It should certainly not be concluded from it that externalism about the process of art production should be restricted to abstract art. Indeed, the core elements at work in the example seem to be ubiquitously present in art and its production. That is, unless one holds a very naïve picture of art as merely expressing some autonomous, determined, pre-existing internal mental reality, one should acknowledge the essential role of material and other external elements in the very possibility of art. It is only through the presence of dyes and walls that depictive art emerged in primitive caves. It is only once the appropriate materials are present that the laborious process of using them to acquire picturing skills can set off. It was the invention of linseed oil paint at the beginning of the fifteenth century which enabled the Flemish primitives to reach an unseen photographic realism, which was simply impossible to accomplish with a fast drying painting medium like tempera. Another example is the range of devices developed in the fifteenth and sixteenth centuries as drawing aids to linear perspective, such as Alberti's transparent grid that was to be placed in front of a scene. Along the path toward the achievement of such skills, many episodes occur in which the concrete, online interaction between material and artist makes a difference — redirects the orientation which the development of art is taking. The emergence of various forms of depiction is constitutively dependent upon the material means of depiction. Mastering new means, learning to interact with those in order to transcend old means, is often what drives art in new directions, in ways illustrated amply elsewhere in this volume (see also Myin 2000).

But one should not focus too much on the concrete material means and tools relied upon in the concrete production process. The artistic objects themselves, once produced, become an environmental element of primordial importance. New art is to a large degree a confrontation with already existing art — a moment in a tradition. This confrontational dialogue of novel or emerging art with existing art *is* an active manipulation of an environment, comparable, even if less concrete, to the manipulation of the cubes in the example we gave above.

Interestingly, it seems that, in contemporary art, this performative aspect of making a move in an unfolding 'art game' has become, in a number of cases, of criterial importance. Some things seem to count as art, not because of any properties they have as material object, but because of the performative value of presenting them as art. Marcel Duchamp's *Fountain*, his famous urinal on a pedestal, is to be appreciated as making a point, rather than as having aesthetic merit as a physical object. This performative aspect of art is of primordial importance to art, and cries out for an externalist understanding. It will reappear repeatedly in the rest of our discussion. Indeed, it will turn out to be essential to understand art appreciation — art consumption rather than art production — to which we now turn.

To what extent can art perception, or art 'consumption', be illuminated by externalism, be it of the active or explanatory kind? Active externalism seems to be particularly well placed to do justice to the concrete material circumstances of perceiving art. Time and again, theorists have emphasized the importance of these. The pure materiality of the artwork itself, as well as the concrete specifics of the perceptual conditions, have been considered to be necessary factors in coming to properly appreciate art. In discussing Rothko paintings, for example, Julia Kindy has remarked that its 'all-encompassing, radiant atmosphere' can never be properly understood without facing the work. 'The scale alone of it', she writes, 'is meant to relate to the body, so that the painting can be "absorbed" by more than the eyes' (Kindy 1999). As Kindy stresses, a reproduction cannot lead to that intense 'physical experience'. The concrete specifics of the situation seem to be required, very much in the way thematized by active externalism (see also Myin 2000).

But there is more potential for externalism. The externalist character of art production, the fact that art creation essentially involves both material things as well as a tradition, the interaction of which it owes its identity to, is a factor covered by explanatory externalism. This importance of the broad context of production reappears as crucial in the domain of consumption, as art appreciation is criterially sensitive to it. We appreciate Caravaggio as a master because of what he did in a tradition. If 'The Incredulity of Saint Thomas' had not been created in 1601, but in 2010, it would be quite differently appreciated. It is no longer possible to create a baroque masterpiece today.

So both the concrete circumstances as well as the broader contextual factors matter in both art production and art consumption, and the importance of both can be accommodated by externalism, with active externalism most appropriate for concrete conditions of production and perception, and explanatory externalism pertaining to the essential role of broader context.

There are two worries that might occur here. First, the concern might again arise that explanatory externalism makes externalism too cheap (a concern already considered in section 1.4). If explanatory dependence on past conditions makes something 'mental' 'partially constituted by the world', externalism about art and aesthetics might win too easily. It might become trivial, as nobody will deny that art and its appreciation depend on the past in some way. The reply we already gave to the more general version of this worry in section 1 applies, *mutatis mutandis*, here as well. The claim that external circumstances are necessary to explain art appreciation is demonstrably not trivial, because rival explanations which refrain from invoking external circumstances in explaining aesthetic appreciation are possible, and actually exist. Take, for example, the programme which can be called 'neuroaesthetics' (Livingstone 1988; Zeki 1999). Though ambitions vary in different theorists, one conspicuous line in this programme is to explain at least some cases of aesthetic appreciation in terms of strictly neural factors. The idea is that there is a significant degree in which the 'what it is like' of experience is a construct of the brain—due to non-stimulus related ways in which the brain is structured, rather than to the stimulus itself. Though again colour is the recurrent example here

(Zeki 1999) of something 'added' by the brain, the scope of potential application is much broader. One such application starts from the theory that the 'visual' brain treats the retinal image in different pathways, dedicated to different aspects such as form, motion or colour. The different properties of these different pathways, such as a difference in resolution between for example the colour and the form pathway can be used to explain various perceptual phenomena. One can bring such explanation to aesthetics, by explaining aesthetic success in terms of exploiting this neural structure. Margaret Livingstone offers such an explanation of Mondrian's Broadway Boogie Woogie. According to Livingstone (1988), the sense of movement elicited by Mondrian is due him cleverly exploiting the fact that the colour, but not the form pathways in the brain, are sensitive to low level luminance contrast. As a result, the colours do not seem firmly confined within spatial borders, and a surprising and pleasing dynamic effect is generated (*ibid*.). Without going into further detail (see Myin 2000 for more as well as additional pointers), one can see that it is plainly possible, if not actually done, to provide explanations of the appreciation of art which do not rely at all on external factors, but which remain within a strictly neural domain. Explanatory externalism regarding art consumption, therefore, is not trivial.

A second worry one might have is that the emphasis on material properties that get emphasized through active externalism, stands in tension to the emphasis on broader contextual factors in explanatory externalism. To the extent that it is its place in the greater drama of art development that is important for a work's status, it might seem that its local material properties become of lesser relevance. A Caravaggio produced today and the real one can be imagined to have exactly the same material properties, yet as we said above ourselves, would have a totally different standing as a piece of art.

The resolution of this worry lies in becoming aware of the fact that there need not be a contrast between the one and the other emphasis. Both aspects matter normally, but in some occasions one of them can be more significant, without there being a general rule which separates these circumstances beforehand. Mistaking the question of 'form' versus 'process' as one which needs to be decided in favour of either the one or the other can lead to a narrow understanding of art and its appreciation. We think exactly this has happened in another internalism/externalism debate, one not in the philosophy of mind, but in analytical aesthetics. We will now turn to this discussion, in order to make the point that the varieties of externalism discussed up to now, in allowing to do both justice to 'form' and 'process', can satisfactorily resolve this debate.

2.2 Internalists versus contextualists in analytical aesthetics

Theories of art in contemporary debates in analytical aesthetics as well can be broadly divided into 'internalist' and 'externalist':[3] according to internalism, the properties that determine the identity and value of artworks are internal to the art object, so that everything relevant to the appreciation of art is directly available in a perceptual encounter with the work; externalism is the opposite position which holds that what makes an object an artwork is external to it, so that its proper appreciation essentially involves contextual background knowledge. The first half of the twentieth century was dominated by the internalist position, endorsed by the traditional aesthetic theories of art. Aesthetic theories attempt to find the essence and value of art in the distinctive types of phenomenal experience artworks elicit on the part of the receiver. Mainly due to recent developments in the arts, externalism in now increasingly gaining upper hand, with the emergence of so-called 'contextualist' theories which maintain that objects acquire their artistic status essentially from art-historical, institutional, or theoretical factors. It is our contention that there is some partial truth in both opposite positions. Traditional aestheticism acknowledges the constitutive role of perceptual experience in art but ignores the historical and socio-cultural dimensions of art that are indispensable for its full appreciation. Contextualism, on the other hand, foregrounds these background dimensions at the expense of attention to the visual properties of artworks. From the active and explanatory externalist perspective treated in the above, an appealing middle way between both extreme positions suggests itself, which acknowledges that a proper appreciation of art goes beyond immediate experience and its particular conditions, but which retains a central role for aesthetic experience in art. That is, the artwork should not be reduced to the mere material 'end product' but is better understood as a complex process of which the product is merely a part; and the aesthetic experience involved in its appreciation is to be understood as an active engagement with the artwork, rather than a private, subjective episode separable from interpretation and cognitive content.

Traditional aesthetic approaches to art have typically aimed to invoke aesthetic experience as to define art. Furthermore, such approaches reduce aesthetic appreciation to what is given through an experiential encounter with the intrinsic properties of a work, excluding facts external to it. Art and the aesthetic are taken to occupy a separate realm, disconnected from other modes of experiencing. According to Edward Bullough (1912) and Jerome Stolnitz (1961), the aesthetic experience occurs when one adopts the appropriate 'aesthetic attitude'. Such an attitude is a special sort of

---

[3] The expressions 'internalism' and 'externalism' in this context are due to Danto (2003, p. xvii).

attention to works of art in which the receiver adopts a state of mind that is 'distanced', 'detached' or 'disinterested'. Monroe Beardsley (1981) appeals to the notion of an 'aesthetic object'. He says that certain properties of the experience are distinctive to our intercourse with such objects. Such experiences have in common that they are 'complex', 'intense' and 'unified'. The aesthetic value of artworks, then, lies in their capacity to elicit experiences of such kind. The formalist approaches of Clive Bell (1914) and Roger Fry (1920) define the aesthetic in terms of 'significant form'. Inspired by the emergence of abstract art and cubism in the beginning of the twentieth century, they insisted that the aesthetic interest in an artwork is completely derived from abstract features of its appearance, such as lines, colours, and spatial organization.

Formalism has been the dominant aesthetics in the heyday of modernism and was championed by the influential New York art critic Clement Greenberg. In his paper 'Modernist Painting' (1960),[4] Greenberg understands modernism as a teleological process in which every art form searches out its own essence, and that essence is to be found in the specificity of every medium. Each art form proceeds through gradually throwing off all norms and conventions that turn out not to be essential to the medium of that art form. Thus the essence of painting, which distinguishes it from for instance sculpture, rests on intrinsic properties such as the flatness of the support and the delimitation of that flatness by the frame.[5] Greenberg understands successive art movements—like impressionism, cubism, and abstract expressionism—in terms of an internal development of the means and the possibilities of painting itself. That development has resulted in a progressive evolution toward 'autonomous' and 'pure' painting, which lacks any reference to the outside world and any association with other art forms. According to Greenberg, modernism reaches its highest point with the post-painterly abstraction of Morris Louis and Kenneth Noland.[6] In their abstract paintings, made on unprepared canvas and deprived of any brushstroke, all depth effects and tactile associations were purged away. The end result of the modernist developmental process is a transformation from representation to the painting *qua* material object itself.

Despite its huge influence, formalistic aesthetics is implausible, being untenably 'internalistic' in two senses. First, it evokes a narrow conception of aesthetic experience as being purely subjective, private and separable from thought and interpretation. Such an idea of 'pure' vision is a framework for vision which we think should be rejected for both philosophical and scientific reasons (Churchland, Ramachandran and Sejnowski 1994;

---

4   Reprinted in O'Brian (1993).
5   See Greenberg's 'After Abstract Expressionism' (1962), reprinted in O'Brian (1993).
6   See 'Louis and Noland' (1962), reprinted in O'Brian (1993).

Findlay and Gilchrist 2003). Second, formalism is excessively preoccupied with the art object and its manifest properties. Artworks are conceived as autonomous aesthetic objects, valued merely for the pleasure of their form. That internalist conception of artwork has definitely proven to be empirically false since the mid-1960s, when the period of modernism has abruptly come to an end. Since then, the art world is in a condition in which apparently *everything* can count as an artwork: white paintings, installations that fill the gallery space with clutter, performances that enact meaningless communication, repetitive films projected on museum walls, etc. A lot of contemporary art looks more like non-art than like previously produced art, so it is unlikely that art can solely be understood in terms of manifest perceptual or formal properties. What distinguishes Carl Andre's tiles from ordinary floor covering, or Tracy Emin's bed from an ordinary bed?[7] The current condition of the art world has been anticipated by Marcel Duchamp's so-called ready-mades, introduced already in the beginning of the twentieth century. The aesthetic merit of Duchamp's already mentioned *Fountain* does not reveal itself by merely examining the physical object. Rather, it is to be properly appreciated as a particular making of a point: the work actually *challenged* the traditional identification of artworks with properties intrinsic to the object. Thus internalism fails, despite its still being the dominant line of thinking within neuroaesthetics. Semir Zeki's (1999) suggestion that artworks reveal an understanding of the role of the early visual brain in the construction of the structure of appearances only accommodates the perceptual experience of abstract works which exploit formal visual elements in relative isolation. As with the traditional aesthetic theories, the attempts to ground aesthetics in an understanding of the neuropsychology of vision evoke some notions of 'pure' vision as opposed to cognition and interpretation and inevitably narrow the scope of the aesthetic to a very restricted class of artworks.

Supported by the challenge to internalism posed by instances of avant-garde art, so-called 'contextualist' approaches have emerged that attempt to define art *without* referring to the manifest, directly perceivable properties of the art object. Contextualists have turned their attention away from intrinsic, directly perceivable properties to extrinsic or relational properties, notably of a social, historical, or 'institutional' nature. The exploration of such relational properties as providing artistic status to works is found in the work of, among others, Arthur Danto (1964), Georges Dickie (1974), Timothy Binkley (1977) and Noël Carroll (1986). According to the contextualist, what makes something an artwork is not its intrinsic properties but

---

[7] In 1967, The American minimalist artist Carl Andre introduced floor pieces consisting of metal tiles, whereby the spectator was invited to walk over them. In 1999, the British artist Tracy Emin exhibited *My Bed* for her nomination of the Turner Prize: her bed that was slept on, malodorous and wrinkled after a week of illness, and with all her personal things she used in it, like books, cigarettes and bottles.

how it relates to a loosely characterized institution or an historically defined socio-cultural practice, which requires the relevant background knowledge on the part of the receiver. The properties that make an artwork into an artwork are 'invisible', external properties, of which the proper appreciative response, then, is primarily a matter of *interpretation* rather than perceptual experience. As a consequence, there is no room anymore for perception, and thus aesthetic experience, to play any constitutive role in the identity and value of art.

Danto insists that understanding art requires attention to features of artworks that are not perceptual at all. He writes: 'To see something as art requires something the eye cannot descry — an atmosphere of artistic theory, a knowledge of art history: an artworld' (1964, p. 581). Danto takes Andy Warhol's Brillo Box — a facsimile of a cardboard box for soap pads, first exhibited in 1964 — to be a paradigm case in support of his bold anti-aestheticism. If Warhol's Brillo Boxes are indistinguishable in appearance from ordinary Brillo Boxes, so the argument goes, how could the former be art and not the latter? Danto anwers that both are differentiated by theory. The Brillo Box acquires its artistic status against the background of the theoretical suppositions endorsed by pop artists. For Danto, the introduction of the Brillo Box meant that 'visuality' had become totally irrelevant for the contemporary visual arts: 'Visuality drops away, as little relevant to the essence of art as beauty proved to have been… Whatever art is, it is no longer something primarily to be looked at' (Danto 1997, p. 16). What artworks mean depends not on their visual or perceptual features but on the historical context, especially the context of an historically situated theory. In a similar vein, Carroll and Binkley argue that the acknowledgment of the existence of 'non-aesthetic' artworks, such as Duchamp's readymades, demonstrates that art is not essentially perceptual.[8]

We agree with the contextualist that many instances of contemporary art falsify traditional aesthetic theories. We disagree, however, with the further inference that such types of artworks are 'anti-aesthetic' or 'non-perceptual'.[9] As we have said, traditional aesthetic theories fail on the basis of two internalist presuppositions, that is, a narrow, sensationalist conception of perceptual experience, and a narrow, because exclusive, focus on the material 'end product' of the artwork. Both presuppositions

---

8 Thus Carroll: 'Now my point against aesthetic theorists of art is that even if *Fountain* does not promote an aesthetic interaction, it does promote an interpretive interaction. Moreover, an interpretive interaction, including one of identifying the dialectical significance of a work in the evolution of art history, is as appropriate and as characteristic a response to art as an aesthetic response' (Carroll 1986, p. 66).

9 See Shelley (2003) for a similar point. Shelley too argues that the position which we call contextualism wrongly accepts the existence of non-perceptual art by assuming a narrow, sensationalist conception of aesthetic experience. He argues that Binkley and Carroll mistakenly attribute such a narrow conception to the aesthetic theories of Frank Sibley and Francis Hutcheson respectively.

are concisely revealed in Beardsley's notion of a 'perceptual object': 'A perceptual object is an object some of whose qualities, at least, are open to direct sensory awareness' (Beardsley 1981, p. 31). *Only if we assume these internalist presuppositions*, the anti-perceptualist claim would follow. Thus Danto's arguments against the 'visuality' of the visual arts strongly depend on both internalist presuppositions. Consider his famous method of indiscernibles: it consists in identifying the perceived identity that is common to a 'mere thing', as he calls it, and an artwork.[10] As such, Danto assumes that one can isolate the interpretative aspect of an artwork as being its essence, existing over and above the physical object and the uninterpreted perception.[11] From this he concludes that it is interpretation, not visual experience, which determines the relationship between an artwork and its material counterpart. This conclusion no longer holds, however, if we reject both internalist presuppositions in the first place.

2.3 *Contemporary art and aesthetic experience*

Traditional aestheticism, notably in its formalist fashion, fails in its insistence upon a sensory conception of experience and upon its exclusive focus on the material object and its formal properties. Greenberg insists that 'visual art should confine itself exclusively to what is given in visual experience, and make no reference to anything given in any other order of experience' (O'Brian 1993, p. 91). For Greenberg, modernist paintings are properly appreciated 'with the eyes' only, against a neutral, white gallery wall, making abstraction of all contextual and social factors, to reveal its specifically visual effects. Note how paradoxal this demand actually is. For paradigmatically modernist paintings in fact *resist* being responded to in such a 'purely' visual manner. Take for instance the paintings of Morris Louis. Louis's so-called 'veil paintings' of the mid-1950s were composed of several diaphanous layers, generated through a sophisticated technique of pouring thinned paint over an unprepared canvas. In these works, form is generated by the direction of the poured paint, drawn by gravity instead of the artist's brushing activity. By pursuing 'flatness' at the extreme, through even eliminating any brushstroke as to eliminate all tactile—'sculptural' and thus not purely 'pictorial'—associations, Louis's paint-

---

10 Danto introduces his method with a thought experiment in which he proposes to imagine several red, square canvasses that are visually identical yet are totally different objects: a painting entitled 'The Israelites Crossing the Red Sea', a psychological portrait entitled 'Kierkegaard's mood', a still life of a red table cloth, a minimalistic painting, a mere nonartistic object, etc. (Danto 1981, pp. 1–3). The point of the thought experiment is that the differences cannot be discovered simply by looking. One needs to have knowledge of the origin of the work, when it is made and by whom, and what it means.

11 Danto writes: 'In the philosophy of art there is no appreciation without interpretation. Interpretation consists in determining the relationship between a work of art and its material counterpart. But since nothing like this is involved with mere objects, aesthetic response to works of art presupposes a cognitive process that response to those mere things does not' (Danto 1981, p. 113).

ings express Greenberg's ideal of pictorial 'purity' and execute his programme of immanent self-criticism.[12]

Ironically, aesthetically appreciating Louis's paintings inevitably demands reference to how they were brought about. We would not understand their contribution to the internal development of modernist painting if we would simply follow Greenberg's recommendations. Yet this does not mean that the visual appearance of the paintings would be irrelevant for their proper appreciation. What this means is that understanding an artwork is not separable from understanding its making. In broad agreement with the process-centred approaches to art offered notably by Dennis Dutton (1979), Gregory Currie (1989), and David Davies (2004), we insist not to restrict the object of appreciation to the properties of the art object but extend it to the artist's *achievement* in producing such an object (*cf.* section 3.1).[13] Dutton stresses that in aesthetic appreciation of a painting, we are confronted with results of human agency. He writes: 'In order to grasp what it is that before us, we must have some notion of what the maker of the object in question has done, including some idea of the limitations, technical and conventional, within which he has worked' (Dutton [1979] 2001, p. 104). We ascribe value to an artwork not merely on the basis of its strictly perceivable properties but on the basis of what the artist achieved. Yet this fact does not diminish the constitutive role of perceptual experience in aesthetic appreciation. It only challenges the narrow conception of perceptual experience as a process that occurs totally independently of thought and interpretation. To perceive is not merely to react to environmental stimuli. It is not conceptually independent from attention, memory, and anticipation. Consider the analogy with hearing speech. I do not first hear something and then, in the light of my knowledge, recognize it as spoken words. Rather, I directly hear speech, not merely physical sounds. This might be extended to the perception in general. It is certainly not the case that I first receive certain stimuli and interpret them after-

---

12   This is what Greenberg writes about Louis: 'The more closely color could be identified with its ground, the freer it would be from the interference of tactile associations; the way to achieve this closer identification was by adapting watercolor technique to oil and using thin paint on an absorbent surface. Louis spills his paint on unsized and unprimed cotton duck canvas, leaving the pigment almost everywhere thin enough, no matter how many different veils of it are superimposed, for the eye to sense the threadedness and wovenness of the fabric underneath' (O'Brian 1993, p. 97).

13   Currie and Davies seem to go further and attempt to *identify* the artwork with the process rather than the object. According to Currie, an artwork is not a physical structure produced, but rather the way in which an artist arrived at that structure. That which we perceive is an instance of the structure the artist arrives at, which allows us to appreciate, in part, the real work, which is the way the artist selected that structure. For Davies too, the real work is the process, or series of actions, by which an artist arrives at his or her product and not just the product itself. According to Davies, '[A]rtworks are *doings* of a particular kind. To appreciate a work is to appreciate what was done' (2004, p. 174). The end product, say, a painted canvas, is merely what Davies calls 'the focus of appreciation' of the work, in which the artist's ideas and efforts are embedded and through which we can come to appreciate the artist's achievement.

wards. And there is no reason why this should not hold for the perception of artworks. We do not first 'see' a patch of colours and shapes and then 'infer' that what we see is an artwork. Rather, we can perceive artwork directly 'as an artwork', that is, as a product of human agency.[14]

With these considerations in mind, we are in a position to challenge the contextualist's claim about the existence of so-called 'anti-aesthetic' art. The emergence of art after modernism in the mid-1960s, which for Danto (1997) famously marked 'the end of art', has been commonly thought to be the crucial test case for contextualism, since it allegedly drove a crucial wedge between 'aesthetic' and 'non-aesthetic' art. Even though we surely acknowledge that an important change has occurred in that period, the change is not as dramatic as commonly assumed. Once the internalist conceptions of perceptual experience and artworks are given up, we have no longer reason to assume a radical discontinuity between aesthetic and non-aesthetic art.

To illuminate this, let us take a closer look at what happened after the end of modernism. From then, it was definitely no longer legitimate to simply speak of artworks as strictly 'material entities'. Art evolved progressively toward a 'dematerializing of the art object',[15] marking a significant shift from 'object' to 'process'. That development emerged primarily as a direct reaction against the hegemony of Greenbergian modernism and a resistance against the autonomy of the artwork and the specificity of the medium. American modernist painting became more and more disconnected from life. As a reaction, the generation of artists that followed brought art back into contact with ordinary life. Artists incorporated 'ordinary' things that previously did not belong to art — products from mass media, industrial materials, everyday appliances, etc. — and made them their medium, material and subject. They produced artworks that were neither paintings nor sculptures, and they explored new media like performance art, installation art and video-installation art.

A contemporary artwork no longer coincides with the 'material object' that hangs on the wall or stands on a pedestal in the gallery or the museum. Bruce Nauman performs certain repetitive acts in his studio and registers them on video; Richard Long makes walks in nature, collects sticks and stones, and exhibits them in the art gallery; Sol LeWitt's wall drawings consist of descriptions of wall paintings, instructions that are

---

14 Joseph Margolis also makes the analogy between hearing speech directly and seeing art directly, as an argument against Danto's indiscernibility method. He writes 'If Danto refused the option of ever perceiving artworks directly as artworks, *he would have to deny* as well that we ever hear speech directly' (Margolis 2000, p. 333); 'Indiscernibility between artwork and non-artwork or between artworks of one kind and artworks of another (the red square cases) *can never be more than occasional and marginal* — and conceptually dependent on the general discernibility of what count as artworks' (*ibid.*, pp. 329–30).

15 As it was termed by the American artist and art theorist Lucy Lippard (1973). See also Matravers (2007) and Davies (2007) for discussions about the dematerialization of the art object.

executed afterwards on a museum wall by assistants.[16] In such cases — instances of respectively performance art, land art and conceptual art — the artwork is a complex whole. It would be mistaken to simply identify the particular wall paintings of LeWitt with the artwork, since they are only some of the many possible executions of the plan. As a conceptual artist, LeWitt wanted to emphasize the priority of the idea, the conception in the mind of the artist, over the execution. If we would ignore the plan and only consider the visual qualities of a particular executed wall painting, we would not have fully understood the work. But neither can the work be identified with the plan alone: it is rather the totality of the particular execution — or some of the subtly different executions — *together* with the plan on which they are based.[17] The dematerialization of the art object radically breaks with the traditional identity of the artwork with its material presence.

Yet, as we have seen, an exclusively object-centred view fails for art *in general*. The crucial difference between traditional and contemporary art may reside primarily in their divergent emphases on 'object' and 'process' respectively, but both aspects remain important for all art nonetheless. Rather than speaking of a difference in 'kind', a radical discontinuity between 'aesthetic' and 'non-aesthetic' art, it would be more accurate to speak of a matter of 'degree' with regard to whether the priority is given to object or process. For how 'dematerialized' contemporary artworks may be, they are *still* properly conceived as objects of visual attention. A full appreciation of them still requires going to the gallery and visually exploring them, rather then to simply read a description of them. Due to their elusive nature, contemporary artworks appeal to visual perception in a different way than traditional artworks: they demand a more active engagement on the part of the perceiver, but that doesn't make them 'non-perceptual'. In fact, if an active externalist approach to perception such as the sensorimotor contingency theory is correct, quite the opposite holds. As Gregory Currie (2007) has insightfully remarked, the difference between traditional and contemporary art can be understood in terms of a difference with respect to *direction* of priority with regards to process or object.[18] The appearance of a Nauman video or Long installation is important in the sense that it engages the spectator in the right way with the action performed by the artist. The role of perceptual experience, here, differs from traditional artworks in the sense that it is directed to a non-perceptual engagement with the work. Conversely, with traditional artworks, the direction is 'from act to experience': 'the act is likely to be conceptualized as one involving a response to certain problems, including technical

---

16 See Archer (2002).
17 See Pillow (2003).
18 Which Currie calls 'action' and 'result' respectively.

problems to do with the manipulation of materials, but also to do with, say, handling of perspective, colour relations, and the inclusion in the work of certain formal relations' (Currie 2007, p. 48).

We might conclude, then, that, *pace* Danto, art after modernism does not mark the end of 'visuality' — and, as a consequence, the end of 'aesthetic experience' — in the arts. Rather than being the ultimate counter-examples to aesthetic approaches *tout court*, contemporary artworks only challenge the assumption that our visual response to visual artworks is 'purely' phenomenal. From the externalist position about the mind discussed here, however, visual perception is primarily an activity. The activity of looking at artworks may serve different purposes and often needs the aid of narrative or theory. In general, it involves seeing them as products of human agency and thus goes beyond the domain of the purely sensory.[19]

References

Adams, F. and K. Aizawa (2001), 'The Bounds of Cognition', *Philosophical Psychology*, **14**: 43–64.
Adams, F. and K. Aizawa (2008), *The Bounds of Cognition*, Malden (MA), Blackwell.
Archer, M. (2002), *Art Since 1960*, London, Thames & Hudson.
Beardsley, M. (1981), *Aesthetics: Problems in the Philosophy of Criticism*, Indianapolis, Hackett.
Bell, C. (1914), *Art*, London, Chatto and Windus.
Binkley, T. (1977), 'Piece: Contra Aesthetics', *The Journal of Aesthetics and Art Criticism*, **35** (3): 265–77.
Bullough, E. (1912), 'Phychical Distance as a Factor in Art and as an Aesthetic Principle', *British Journal of Psychology*, **5**: 87–98.
Chambers, D. and D. Reisberg (1985), 'Can Mental Images be Ambiguous?', *Journal of Experimental Psychology: Human Perception and Performance*, **2**: 317–28.
Carroll, N. (1986), 'Art and Interaction', *The Journal of Aesthetics and Art Criticism*, **45**: 57–68.
Chemero, A. (2010), *Radical Embodied Cognitive Science*, Cambridge (MA), MIT Press.
Churchland, P.S., V.S. Ramachandran and T.J. Sejnowski (1994), 'A Critique of Pure Vision' in C. Koch and J.L. Davies, Eds., *Large-Scale Neuronal Theories of the Brain*, Cambridge (MA), MIT Press.
Clark, A. (2003), *Natural-Born Cyborgs: Minds, Technologies, and the Future of Human Intelligence*, New York, Oxford University Press.
Clark, A. and D. Chalmers (1998), 'The Extended Mind', *Analysis*, **58** (1): 7–19.
Currie, G. (1989), *An Ontology of Art*, New York, St. Martin's Press.
Currie, G. (2007), 'Visual Conceptual Art' in P. Goldie and E. Schellekens, Eds., *Philosophy and Conceptual Art*, Oxford, Oxford University Press.
Danto, A. (1964), 'The Artworld', *Journal of Philosophy*, **61**: 571–84.
Danto, A. (1981), *The Transfiguration of the Commonplace: A Philosophy of Art*, Cambridge (MA), Harvard University Press.
Danto, A. (1997), *After the End of Art: Contemporary Art and the Pale of History*, Princeton, Princeton University Press.
Danto, A. (2003), *The Abuse of Beauty: Aesthetics and the Concept of Art*, Chicago and La Salle (IL), Open Court.
Davies, D. (2004), *Art as Performance*, London, Blackwell.

---

19 The research by the authors was supported by the BOF-project 'Visual Imagery as Perceptual Activity', funded by the University of Antwerp.

Davies, D. (2007), 'Telling Pictures: The Place of the Narrative in Late Modern "Visual Art"' in P. Goldie and E. Schellekens, Eds., *Philosophy and Conceptual Art*, Oxford, Oxford University Press.

Dickie, G. (1974), *Art and the Aesthetic: An Institutional Analysis*, Ithaca (NY), Cornell University Press.

Dutton, D. (1979), 'Artistic Crimes: The Problem of Forgery in the Arts', *The British Journal of Aesthetics*, **19** (4): 302–41. Reprinted in A. Neill and A. Ridley, Eds., (2001), *Arguing About Art, Second Edition*, New York, McGraw-Hill.

Findlay, J.M. and I.D. Gilchrist (2003), *Active Vision: The Psychology of Looking and Seeing*, Oxford, Oxford University Press.

Fodor, J.A. (1980), 'Methodological Solipsism Considered as a Research Strategy in Cognitive Psychology', *The Behavioral and Brain Sciences*, **3**: 63–109.

Fry, R. (1920), *Vision and Design*, London, Chatto & Windus.

Gibson, J. (1966), *The Senses Considered as Perceptual Systems*, Boston, Houghton-Mifflin.

Hardin, C.L. (1988), *Color for Philosophers*, Indianapolis and Cambridge: Hackett.

Hurley S. (1998), *Consciousness in Action*, Cambridge (MA), Harvard University Press.

Hurley, S. (2010), 'The Varieties of Externalism' in R. Menary, Ed., *The Extended Mind*, Cambridge (MA), MIT Press.

Hurley, S. and A. Noë (2003), 'Neural Plasticity and Consciousness', *Biology and Philosophy*, **18**: 131–68.

Kindy, J. (1999), 'Of Time and Beauty', *Journal of Consciousness Studies*, **6** (6–7): 61–3.

Land, M.F. (2004), 'Eye Movements in Daily Life' in L.M. Chalupa and J.S. Werner, Eds., *The Visual Neurosciences* (Vol. 2), Cambridge (MA), MIT Press.

Lippard, L. (1973), *Six Years: The Dematerialization of the Art Object from 1966 to 1972*, New York, Praeger.

Livingstone, M. (1988), 'Art, Illusion and the Visual System', *Scientific American*, **256**: 78–85.

Margolis, J. (2000), *The Arts and the Definition of the Human: Toward a Philosophical Anthropology*, Stanford (CA), Stanford University Press.

Matravers, D. (2007), 'The Dematerialization of the Object' in P. Goldie and E. Schellekens, Eds., *Philosophy and Conceptual Art*, Oxford, Oxford University Press.

Menary, R. (2006), 'Attacking the Bounds of Cognition', *Philosophical Psychology*, **19**: 329–34.

Menary, R. (2007), *Cognitive Integration: Mind and Cognition Unbounded*, Basingstoke, Palgrave.

Menary, R. (2010), 'Introduction: The Extended Mind in Focus' in R. Menary, Ed., *The Extended Mind*, Cambridge (MA), MIT Press.

Myin, E. (2000), 'Two Sciences of Perception and Visual Art', *Journal of Consciousness Studies*, **7** (8–9): 47–59.

Myin, E. and K. O'Regan (2009), 'Situated Perception and Sensation in Vision and Other Modalities: A Sensorimotor Approach' in M. Aydede and P. Robbins, Eds., *The Cambridge Handbook of Situated Cognition*, Cambridge: Cambridge University Press.

O'Brian, J., Ed. (1993), *Clement Greenberg: The Collected Essays and Criticism*, (Vol 4), Chicago and London, The University of Chicago Press.

O'Regan, K. and A. Noë (2001), 'A Sensorimotor Account of Vision and Visual Consciousness', *Behavioral and Brain Sciences*, **24** (5): 939–73.

Pillow, K. (2003), 'Did Goodman's Distinction Survive LeWitt?', *The Journal of Aesthetics and Art Criticism*, **63**: 365–81.

Rowlands, M. (2009), 'Enactivism and the Extended Mind', *Topoi*, **28**: 53–62.

Ross, D. and J. Ladyman (2010), 'The Alleged Coupling-Constitution Fallacy and the Mature Sciences' in R. Menary, Ed., *The Extended Mind*, Cambridge (MA), MIT Press.

Rupert, R. (2004), 'Challenges to the Hypothesis of Extended Cognition', *Journal of Philosophy*, **101**: 389–428.

Rupert, R. (2010), 'Representation in Extended Cognitive Systems: Does the Scaffolding of Language Extend the Mind?' in R. Menary, Ed., *The Extended Mind*, Cambridge (MA), MIT Press.

Shelley, J. (2003), 'The Problem of Non-Perceptual Art', *The Journal of Aesthetics and Art Criticism*, **43** (4): 363–78.

Sutton, J. (2010), 'Exograms and Interdisciplinarity: History, the Extended Mind, and the Civilizing Process' in R. Menary, Ed., *The Extended Mind*, Cambridge (MA), MIT Press.

Stolnitz, J. (1961), 'On the Origins of "Aesthetic Disinterestedness"', *Journal of Aesthetics and Art Criticism*, **20**: 131–44.

van Leeuwen, C., I.M. Verstijnen and P. Hekkert (1999), 'Common Unconscious Dynamics Underly Uncommon Conscious Effect: A Case Study in the Iterative Nature of Perception and Creation' in J.S. Jordan, Ed., *Modeling Consciousness Across the Disciplines*, Lanham (MD), University Press of America.

Wilson, R.A. (2004), *Boundaries of the Mind: The Individual in the Fragile Sciences*, Cambridge, Cambridge University Press.

Wilson, R.A. (2010), 'Meaning Making and the Mind of the Externalist' in R. Menary, Ed., *The Extended Mind*, Cambridge (MA), MIT Press.

Zeki, S. (1999), *Inner Vision*, New York, Oxford University Press.

Joel Krueger

Enacting Musical Content

According to the enactive view, perception is intrinsically active (Hurley 2001; Noë 2004; O'Regan and Noë 2001; Thompson 2005). Perceptual experience acquires content—objects of experience are brought to phenomenal presence—via the possession and deployment of bodily skills: the ability to use our body to explore, manipulate, and engage with the world and things in it. Accordingly, enactive approaches tend to be pitched as theories of *access* (e.g. Noë 2009): accounts of how the possession, deployment, and understanding of bodily skills, as well as the sensorimotor regularities governing our engagement with the world, determine the character (the *how*) and content (the *what*) of perceptual experience—that is, the form of our experiential access to the world. Enactive views take as their target classical computational theories of vision which, according to the enactivist, fail to account for the crucial contribution that agency makes to visual experience (Gangopadhyay and Kiverstein 2009). Though not without their critics, enactive approaches are increasingly influential. Yet proponents have said little of how the account might be extended to sensory modalities other than vision and touch.

This chapter offers the beginning of an enactive account of auditory experience—particularly the experience of listening sensitively to music. It investigates how sensorimotor regularities grant perceptual access to music *qua* music. Two specific claims are defended: (1) music manifests experientially as having complex spatial content; (2) sensorimotor regularities constrain this content. Musical content is thus brought to phenomenal presence by bodily exploring structural features of music. We enact musical content.

1 Preliminaries

Why is this discussion philosophically interesting? First, this view extends enactive accounts of perception beyond the realm of vision and touch. Most discussions of perception within enactive literature take vision as the paradigm case of perceiving. Yet the core claims animating the enactive view apply equally well to auditory perception, I suggest, including the experience of listening to music. Moreover, they allow for an adequate phenomenological portrayal of our sensitive listening episodes—instances of what I will term 'deep listening'—one which captures the experiential features unique to musical experience. Finally, these enactive claims receive robust empirical support from several strands of empirical studies on auditory and music perception, discussed below.

Second, the enactive view here developed challenges dominant tendency in philosophy and psychology to treat musical experience as a relatively passive affair—that is, as consisting of a linear causal process leading from musical piece to listener. John Sloboda has labelled this the 'pharmaceutical model' of musical experience (Sloboda 2005). According to Sloboda, this model rests on the assumption that music's psychological efficacy is much like taking medication. There are pre-determined, consistent effects that result from specific musical structures interacting with specific brain regions (*ibid.*, p. 319). But like Sloboda, I will argue that this 'passive listener model' of music listening ignores the richly interactive nature of musical experience. Musical experience is a form of active perceptual exploration—an active, world-engaged *musicking* (Small 1998).[1] The enactive view here defended thus emphasizes the dynamic and agentive nature of music perception, urging that the embodied and situated listener has a central role in shaping both the character and content of musical experience. Musical experience, like perceptual consciousness more generally, is *transactional* (Putnam 1999).

2 The space of hearing

Do we auditorily experience spatial features of sounds? Brian O'Shaughnessy states flatly that 'we absolutely never perceive sounds to be *at* any place. (Inference from auditory data being another thing.)' (O'Shaughnessy 1984, p. 199). This is because sounds have aspatial phenomenology. O'Shaughnessy doesn't deny, of course, that we can *locate* sounds within the surrounding environment. However, this locational aspect of sound perception is determined by *acoustic* features of the sounds: for example, pitch, timbre, or loudness, such as recognizing that the rising loudness of an ambulance's siren means that it is spatially closer to me than it was ten seconds ago. But again, there is no spatial content intrinsic to the phenomenology of siren hearing on its own. All we hear, rather, is the sound itself. We don't hear the sound as standing in any relation to the space it occupies

---

[1] F. Joseph Smith observes similarly that, '[t]he image of the extended hand, significative of intentionality, connotes the activity required to constitute the proper receptivity. Receiving is, therefore, not inert passivity. The hand must close on the gift, else it will fall to the ground. Similarly, in order to listen to either spoken language or to musical sound, we have to "lend our ears." Listening is not merely passive reaction; it is its own activity' (Smith 1979, pp. 100–111).

(Nudds 2001, pp. 213–4). This is because 'the sound I hear is where I am when I hear it' (O'Shaughnessy 1984, p. 199). O'Shaughnessy thus endorses a 'proximal' theory of sounds according to which sounds are located where their hearer is (Casati and Dokic 2009). When we locate sounds in space, the auditory data present to our ears is 'augmented by mental factors leading to one's hearing the sounds to be *coming from* a specific site' (O'Shaughnessy 2009, p. 125). We infer or work out—again, via acoustic features of sounds or by extra-auditory data from experiences in other modalities (e.g. vision) that do possess genuine spatial content—where the sound sources are likely located. But the sounds-as-heard have no intrinsic spatial phenomenology. Rather, audition plays a supplementary role. According to O'Shaughnessy, it provides inference cues to spatial features and locations that come to us through other sense modalities. However, lacking non-derived spatial content, audition in this way stands in contrast to experiences of other sense modalities.

There are some problems with this view salient to this paper's core concerns. First, a simple phenomenological objection is that the characterization of hearing at work here misdescribes the experiential character of everyday auditory experience. For, we surely don't have the extra-auditory *experience* of inferring or working out the direction and location of sounds. Rather, we hear the sounds themselves, immediately and non-inferentially, as egocentrically located in our surrounding environment. For instance, I immediately hear the laughter and voices of my neighbour's children playing outside *slightly behind me and to the left*, just outside the window of my office; and if I suddenly hear a dull thud followed by crying—perhaps one of the children has run into the front yard and tripped over the sprinkler along the way—I know exactly where to go to offer aid. Likewise, I hear the bird squawking loudly as he passes by *directly above me*; the voices of the couple arguing emanate from the apartment *right next to mine*. When someone calls my name, I immediately turn without having to think about it in the direction of the sound; and blind people, too, are quite capable of tracking and responding to sound events despite the absence of extra-auditory visual input (Hamilton 2009, p. 176). Phenomenologically speaking, it appears that auditory experiences are locational. They represent both *what* is happening (e.g. children playing outside) as well as *how* what is happening stands in relation to oneself (e.g. slightly behind me and to the left).

Moreover, our ability to non-inferentially locate sounds in egocentric space purely by hearing them has important behavioural consequences. If, while attending a baseball game and chatting up a friend in the stands, I suddenly hear the crack of a bat and the whizz of an oncoming baseball, I know immediately which way to duck without having to first consider extra-auditory data (which in this case is likely not even available to me, given that I'm looking away from the sound source). The ability to track and quickly respond to events in the environment rests on having immediate phenomenal access to the spatial location carried by sounds. A step-wise process depending upon the access and utilization of extra-auditory data, or necessarily mediated by an inferential 'working out' of a sound's location, places an unnecessarily excessive computational burden on the perceiver—and thus would, accordingly, significantly impede their reaction times. O'Shaughnessy's view is thus phenomenologically implausible (it is also challenged by empirical research, as will become apparent throughout this paper). And while this phenomenological objection is not in itself a devastating refutation, the view's descriptive implausibility should at least give us some pause.

Another difficulty with the view, as Casey O'Callaghan notes, is the paucity of egocentric information provided by purely *acoustic* features of sounds, such as pitch, timbre, and loudness (O'Callaghan 2010, p. 133). According to O'Shaughnessy, acoustic features of sounds are the primary bearers of whatever spatial data is available from sounds. Again, we hear the rising loudness of a siren and work out from this loudness that it is moving closer to us. But acoustic features seem ill-suited to fill this role. For 'perceptible qualitative attributes, such as pitch, timbre, and loudness fail to correspond reliably to egocentric location, and variations in qualities do not correspond reliably to changes in egocentric location' (O'Callaghan 2009, p. 133). So, pitch and timbre preserve their character independently of spatial location: a musical instrument may exhibit the same pitch and timbre whether it's right next to the hearer or across the street. Similarly, while loudness is perhaps a more reliable indicator of location (e.g. the loud voice on the train is coming from the man in the seat immediately behind mine), there are no fixed sensorimotor rules governing the relation between loudness and distance. A musical instrument playing softly may be right next to the hearer while a loudly playing one is across the street. Some researchers have suggested that loudness constancy may not depend upon, or even be related to, source distance perception at all. Zahorik and Wightman conducted several experiments in which listeners reported robust loudness constancy even when the source distance was varied (Zahorik and Wightman 2001). Listeners also systematically overestimated close-range sources and underestimated long-range source distances (*ibid.*, p. 81). Given the facility with which we track and respond to sound-events in our environment—such as the baseball example discussed above—it is therefore unlikely that this facility turns on our responsiveness to purely acoustic features which, as we've just seen, have a relatively tenuous correspondence relation to the listener's egocentric location. It is rather more likely that sounds themselves bear spatial content.

Finally, one more line of empirically-informed support for this idea comes from work on the neural representation of 'auditory space maps': a neural map of how received auditory information is situated in the surrounding environment (Hyde and Knudsen 2002). It appears that '[a]uditory space maps can be generated without visual input, but their precision and topography depend on visual experience. So, for example, owls raised as if they were blind end up with abnormal, or even partially inverted, auditory maps' (Carr 2002, p. 30). While visual input provides more reliable and topographically organized information—and thus can refine and enhance the auditory information represented by auditory space maps—evidence nevertheless suggests that spatial content cuts across both visual and auditory perception (not to mention other modalities). It is thus a mistake to think of auditory experience as bearing only derived spatial content.

In sum: listening to sounds is an exploration of our world—including spatial and locational aspects of things in it. Sounds routinely furnish spatial information about our world, and we use our auditory experiences to explore and skillfully respond to things happenings in it. Phenomenologically, sounds are thus spatially structured.

3 Exploring musical space

One of the things we quite often hear in our world is music, both live and recorded. I now want to argue that the phenomenology of musical experience is determined by the experience of two forms of spatiality: what I term, respectively, 'inner' and 'outer' musical space. I will argue that, in episodes of 'deep listening'—listening in a voluntary mode of sustained perceptual focus and affective sensitivity, as opposed to hearing music with 'one ear' as a piece drifts idly by in the background—listeners enact the experiential fusing of these two forms of spatiality.[2] Put otherwise, the spatiality of musical structure—and in particular, structural features like *textural qualities* and the temporal regularities of *sonic patterns* (both melodic and rhythmic)—presents music as having an exploratory profile affording this sort of deep listening. My approach in this section and the next one is to offer a descriptive phenomenology of how we enact the spatial content of the form of musical experience I am calling 'deep listening'. The section thereafter then discusses some empirical research that seems to support this phenomenological description.

Experience, according to the enactivist, is always an active encounter with hidden complexity (Noë 2009, p. 473). Conscious phenomena harbour potentially attended-to aspects that invite further exploration. In this

---

2 Composer Pauline Oliveros coined the expression 'deep listening' to refer to her own brand of music composition and listening practices. See http://deeplistening.org. See also Becker (2004).

case of visual perception, the anticipation of how a visually conscious object changes relative to bodily movements disclosing previously-hidden aspects of the object (e.g. leaning to the left or right to bring unseen bits of a solid object into view) is a crucial part of actively perceiving the world. These anticipations or expectations are a form of sensorimotor knowledge—an understanding of how our perceptual relation to the world is mediated by contingent relations coupling bodily movement and sensory change. As noted above, most enactive literature focuses on visual perception as the paradigm case of enacting experiential content. But thinking of experience as an active encounter with hidden complexity is no less true for audition and music perception than it is for other forms of experience. Music, in particular, invites sensitive perceptual inspection. It solicits inspection of discrete constituent units that can be individually attended to, differentiated, and sonically explored.[3] This is because musical structure consists of sonic units extended in time—unlike units of visual objects, which are extended in space—and which are therefore perceptually individuated in virtue of pitch and other temporal characteristics (O'Callaghan 2009). This dynamic temporal structure presents an especially rich sort of exploratory profile.

Temporality is essential both to the structure of musical experience as well as to the exploratory profile a musical piece presents. As Søren Kierkegaard notes, 'aside from language, music is the only medium that addresses itself to the ear… Language has time as its element; all other media have space as their element. Music is the only other one that takes place in time' (Kierkegaard 1959, pp. 66, 67).[4] Of course, other media are also situated in time and can, accordingly, exhibit various changes that betray this temporality. Colours on a painting gradually fade as it ages; shadows pass across the surface of a sculpture, giving it distinctive appearances in the light of early morning versus the dim hues of evening. Kierkegaard's point, however, is that with music—unique among the arts—'temporality is not a matter of "subjectivity" but a matter of the way the phenomenon presents itself' (Ihde 2007, p. 94). I can perceptually explore a painting or a statue—I can sit and gaze intently; walk up to it, move away, tilt my head and look from another angle; touch it and run my hand over its surface, etc.—and in this way become aware of its nature as a temporally situated object. But this movement marks a shift toward the

---

3   Marilyn Nonken observes that, '[d]esigned intentionally for sensory exploration, the musical environment is characterized by the presence of not only harmony and rhythm but also such factors as silence, timbre (instrumentation), dynamic (amplitude), density, texture, gestural and motivic figures, patterns, and audible processes of accretion and degradation (such as *crescendo* or *ritardandi*, the processes of getting louder or slower)' (Nonken 2008, p. 294).

4   Schopenhauer puts the point more strongly when he insists that music is perceived 'in and through time alone, with the absolute exclusion of space' (Schopenhauer 1966, p. 266). It is precisely this view that I will challenge, arguing that both temporality and spatiality are integral parts of music and musical experience.

noetic phenomena—that is, my consciousness *as aware* of the passing of time within my exploratory activity (Ihde 2007, p. 94). Temporality is not immediately manifest within the object *itself* (e.g. the painting as *noematic* correlate) but rather within my intentional relation *to* the object. On the other hand, there seems to be a unique structural intimacy between music and temporality; the latter is an essential part of the former's make-up. We therefore cannot hear music without simultaneously hearing how time is embodied within the music.

Yet by focusing exclusively on the temporality of musical experience, there is a danger of losing phenomenological grip on its inherently spatial qualities. A central feature of music's exploratory profile is space. For, as Robert Morgan notes, 'it would seem to be impossible to talk about music at all without invoking spatial notions of one kind or another' (Morgan 1980, p. 527).[5] To perceive music is thus to perceive space. Though I argued above that all auditory experience bears spatial content, musical experience is unique, I suggest, in terms of the *complexity* of its spatial content. Moreover, with respect to other forms of art, the space of music is perceived precisely *in* its temporality in a way not the case with other art forms.

To begin to get a sense of how this is so, consider first some apt remarks by Merleau-Ponty. Despite a relative lack of interest in music in his writing, Merleau-Ponty nevertheless offers a few quotes of interest. First, like Kierkegaard, he emphasizes the importance of temporality in perceiving music. A piece of music, he says,

> ...comes very close to being no more than a medley of sound sensations: from among these sounds we discern the appearance of a phrase and, as phrase follows phrase, a whole and, finally, as Proust put it, a world. This world exists in the universe of possible music, whether in the district of Debussy or the kingdom of Bach. (Merleau-Ponty 2004, p. 99)

The implication seems to be that the 'world-making' power of music only becomes apparent through the active exploration of a piece—that is, careful attentiveness to the temporal dynamics of a piece of music (i.e. 'as phrase follows phrase') that gradually erect a sonic topography inviting further exploration. The temporal unfolding of music is the movement that begins to open up a piece's inner sonic space (I clarify what I mean by this in a moment).

Elsewhere, however, we find a more substantive phenomenological observation:

---

[5] Morgan writes further that, 'Thus in discussing even the most elementary aspects of pitch organization—and among the musical elements, only pitch, we should remember, is uniquely musical—one finds it necessary to rely upon such spatially oriented oppositions as "up and down," "high and low," "small and large," (in regard to intervallic "distances"), and so on' (Morgan 1980, p. 527).

> When, in the concert hall, I open my eyes, and visible space seems to me cramped compared to that other space through which, a moment ago, the music was being unfolded, and even if I keep my eyes open as the piece is being played, I have the impression that the music is not really contained within this circumscribed and unimpressive space. It brings a new dimension stealing through visible space, and in this it surges forward… (Merleau-Ponty 2002, pp. 257–258)

There are several points of interest in this short passage. Salient to present concerns is the claim that musical experience, while temporal, is additionally infused with representations of space. Music both consumes as well as creates space. Specifically, we can say that Merleau-Ponty differentiates between what we might term 'inner' and 'outer' musical space. The former refers to the space internal to the piece of music itself. It is what we might term *structural* space: that is, the piece's inner syntactical structure established by the way that constituent components (e.g. tones, rhythmic progressions, etc.) hang together, lending the musical piece its sonic coherence as a composed object. This form of musical space is experientially fluid; it can swell and expand, as when a piece of music seems to fill a room and surround us, occupying 'a new dimension stealing through visible space'. 'Outer' musical space, on the other hand — what Merleau-Ponty calls 'visible space' — might also be termed *locational* space. This is the egocentric spatial character of music as locationally perceived (e.g. music heard as coming from the speakers *in front of me* or the stage *to my right*), as something inhabiting a determinate location relative to my bodily orientation.

When we perceive a piece of music, we tend to automatically perceive the piece's inner spatial configuration. This is what it means to listen to music understandingly, to hear it as something with an inner complexity offering up an exploratory profile inviting attentive inspection.[6] And we also tend to have a reasonably clear sense of where the musical source is located spatially, such as when we walk into an unfamiliar apartment for the first time and immediately recognize that the stereo is playing in the next room. However, what I want to suggest is that within deep listening episodes, we enact an experiential fusing of these two forms of musical spatiality such that neither takes phenomenological precedence over the other. Rather, they come together and, in their fusing, open up experiential character of the piece in a new and previously unheard way. This is what gives these episodes their unique phenomenal character. And the animate body, as we will see, plays a central role in facilitating this musical-spatial enaction.

Again, 'deep listening', as I'm using this expression, is a voluntary form of musical experience consisting of sustained attentional focus and affec-

---

[6] The phenomenological consequences of failing to perceive the inner space of music will be discussed in more detail below.

tive sensitivity. It is an immersive form of listening in which the subject *selectively orients* herself to a piece of music by actively attending to its various sound features and their interrelationships—while simultaneously maintaining a state of affective receptivity, or a readiness-to-be-moved, by what is happening sonically in the music. Deep listening is thus a transactive mode of listening involving 'processes such as *exploring, selecting, modifying,* and *focusing of attention*' (Reybrouck 2005, p. 252). Moreover, this deep engagement can have the temporary effect of weakening or obliterating the felt senses of inner and outer. This is the cultivation of an auditory *field state*: an expanded, phenomenally 'full' mode of listening in which focal attention stretches to the very boundaries of the sound as present (Ihde 2007, p. 102). In other words, instead of remaining remotely situated, the deep listener instead has the felt sense of *inhabiting* the sound field, leading to a heightened emotional and affective responsiveness to the musical situation (Vastfjall 2003).

Deep listening is thus sensually richer than involuntary or passive modes of hearing such as being faintly aware of background music playing in a grocery store or restaurant, or hearing the sound of a radio drifting out of a nearby open window. In these latter cases, the *locational* spatial character of a musical piece remains experientially prioritized. For example, we hear unobtrusive *muzak* trickle quietly from speakers above us, staggered across the ceiling of the grocery store; we cringe slightly at the shrill sound of a teenager in the bus seat behind ours listening to hip-hop via the underpowered speaker of his mobile phone; the sound of a radio is momentarily present before slowly diminishing and trailing off as we walk by an open window in a nearby apartment complex. However, since our attention is largely focused elsewhere within these shallow listening episodes (e.g. navigating our shopping cart toward the exit; peering out the window to see if our stop is coming up; hurrying on to make the appointment for which we're already late), the inner or structural space of the piece fails to present itself within any sort of phenomenal immediacy. It is experientially present—again, to perceive a particular auditory event *qua* music is to perceive its inner space, however dimly—but it remains diminished and nonfocal, confined instead to the relative margins of our awareness. This passive hearing is thus a minimally active form of musical engagement due to our lack of attentive inspection and affective engagement. Once more, the outer spatiality of the piece is given phenomenological priority within this mode of hearing; it remains predominantly allocentric in character (i.e. 'The sound of the radio is *over there*' — Turner *et al.* 2007).[7]

---

[7] This is not to suggest that there isn't an egocentric aspect to allocentric hearing. For the 'over there' of heard music is the disclosure of an 'over there' specified as such only in relation to my bodily *here*. So, allocentric and egocentric information within audition are experientially

4 Solitary and social forms of deep listening

I've suggested that the experiential character, as well as the representation of the relation between inner and outer musical space, differs in deep listening episodes. While perhaps a somewhat rarified occurrence, deep listening can nevertheless be enacted within nearly any context where music is carefully attended to (and, indeed, the auditory conditions are sufficiently adequate). It can be a solitary undertaking or, in some contexts, a social affair (more on that below). Importantly, it is an intentional modification of everyday hearing—a kind of playing with perception, so to speak—that emphasizes the unique agility and, indeed, *plasticity* of audition. Don Ihde has helpfully observed that the spatiality of the auditory field exhibits a 'double dimensionality': it simultaneously exhibits both *surroundability* (i.e. an atmospheric or enveloping quality) and *directionality* (i.e. a situatedness or locality) (Ihde 2007, p. 77). This double dimensionality is both a source of modality-specific ambiguity as well as a richness that 'subtly pervades the auditory dimension of experience' (*ibid.*, p. 77). This ambiguity—and indeed, richness—is vividly highlighted within deep listening episodes.

To enact a deep listening episode is precisely to play with the ambiguity at the heart of the musical domain's double dimensionality. The initial phase of deep listening might begin by narrowing one's focal attention to capture the sonic shape and texture of a particular sound feature. This initial gesture inaugurates entry into the inner structural space of the music; it is a focused entry into the piece's temporal dynamics. For example, while listening to a favourite piece, the listener may start by attentively following the contour of a melody, listening to and then eventually 'past' its dynamics as it gradually traces a narrative path through musical time and tonal space. This latter notion serves as 'a designation that corresponds to our perception of music as moving *through* something—for example, from a higher position to a lower one' (Morgan 1980, p. 528). Within deep listening, a melody is experienced as unfolding within a spatialized auditory dimension that the deep listener simultaneously moves to inhabit (hence, the experience's immersive character).

As this listening becomes intensified and further focalized—i.e. the listener listens 'past' the pleasant affective solicitations of the melodic contour and becomes aware of things happening behind or below the melody—a more holistic global sensitivity emerges. Another region of inner space becomes phenomenally accessible: the space of musical *relationships*. The experience as of a melody unfolding within tonal space invites the deep listener to become attuned to perceptual differentiations

---

co-given. But co-givenness is not equivalent to sameness of experiential intensity. The allocentric information of 'shallow' hearing remains experientially focalized; it stands out in a way not the case in deep listening, as the descriptions below will attempt to make clear.

between other things happening in the piece. In other words, this felt appreciation of tonal space amplifies an appreciation of the piece's *texture*—that is, its *density* as comprised of multiple simultaneous sound events. For, attentively focusing on the dynamics of a melody (or other discrete sound units) does not obscure other musical events happening in the piece. Rather, melody and accompaniment, for example, 'do not simply merge into a single temporal continuum but appear to occupy different spatial locations, thus maintaining both individuality and a clear mutual relationship' (*ibid.*, p. 529). So, within this next phase of deep listening, the melody is perceived to unfold within a different region of auditory space than its rhythmic accompaniment; the latter undergirds the former. Both thus trace parallel but distinct paths through tonal space. And in perceiving this distinction, the listener thus becomes acutely aware of the (auditory) spatial relation between melody and rhythmic accompaniment—the relation itself becomes a positive feature of the listener's awareness—further deepening and refining the listener's perception of both melody and rhythmic accompaniment. Put differently, the phenomenal appreciation of the *relation* simultaneously enriches the appreciation of the *relata*. The experience of the piece is thus qualitatively deepened. What began as an attentive inspection of melody has thus gradually shifted to a more subtle appreciation of the inner architectonics of the piece as a whole—an appreciation of the 'aggregate quality' of the various inter-relationships linking musical events together within tonal space (*ibid.*, p. 529). And this subtle shift in quality of attention signals an experiential fusing of inner and outer musical space, a blending of surroundability and directionality. The piece is now inhabited. The listener is in a position to actively explore different aspects of this nested acoustic environment from an inside-out perspective, as it were.

The animate body plays an important role in enacting this sort of musical spatial fusing. This is because bodily movements such as gently swaying back and forth, bobbing one's head, tapping fingers and toes, and of course dancing—more on this in a moment—modulate our perception of the spatial content of musical experience by modulating our relation to different features of the music, such as metre and melody. Bodily gestures in response to musical events can act as a kind of attentional focusing: the animate body, by interactively engaging with the piece, becomes a vehicle for voluntarily drawing out certain features of the piece, such as rhythmic beats or the progression of a melodic contour, by foregrounding them in our attentional field. This 'drawing out' is an enactive and exploratory gesture in response to felt affordances within the music.[8] The listener per-

---

8 There is much empirical evidence indicating that we seem to perceive music as affording various forms of bodily interaction from birth. For discussion of the notion of 'musical affordances', see Krueger (forthcoming a).

ceives the inner space of the piece as a space that can be entered into, experientially, and by doing just this shapes how the experiential content of the piece-as-given becomes phenomenally manifest. We thus hear what the body feels (Philips-Silver and Trainor 2007). And what the body feels are sensorimotor contingencies—possibilities for rhythmic interaction and perceptual exploration that determine the character and content of musical experience.

This idea is reinforced by considering shared episodes of deep listening. Consider, for example, how the shared attentional framework in a live music setting affects the group's mutual perception and appropriation of the music. In particular, consider the role that the crowd's latching onto musical textures and sonic patterns plays in shaping the shared experience of music within a live setting. For example, the simple act of a guitarist casually strumming the first few chords of a popular song—especially at the beginning of a concert—immediately elicits a thunderous roar of approval from the audience. Within a moment, the crowd's attention is galvanized around these textures, snapping into a mode of taut expectancy; the atmosphere is flush with anticipation of the song that will soon follow. Once this familiar refrain begins, the texture (and thus the inner space) of a piece is progressively structured by the introduction of new sound elements: the initial guitar strumming is soon girded by the rhythmic pulsing of the bass accompaniment; keyboards emerge to fill in the sound even more, enriching and deepening the sonic structure; next, the drums enter, stabilizing and accelerating the song's forward momentum; finally, vocals materialize, their aurally-discernible human character lending a sense of qualitative unity to the piece as a whole (in addition to whatever narrative dressing the lyrics offer). This gradually-unfurling sonic world invites shared exploration and appropriation. That this is so becomes clear when we observe how the emergence of each new aspect of these textures elicits new bodily responses from the audience (modulations of head bobbing, swaying, and other whole-body movements, dancing, shouts of encouragement, etc.), as well as a collective refocusing of attention on each emergent sound feature as it comes forward. The musicians perceive and respond to these cues—altering their performance accordingly—which in turn subsequently shapes the audience's further responses and attentional refocusing. Within this organic performer-perceiver interplay, then, a shared attentional framework emerges unique to that time and place. The material mediation of this particular music event—the way that the live music event is embodied in things like the number of listeners, the spatial location of the performance, the musical skills of the performers and audience, and the social values of the attendees—determines both (1) what sort of shared attentional framework emerges in that context, and (2) how musical textures are perceived and appropriated by multiple perceivers via this attentional framework. If the audience were to suddenly disappear, save for one lone listener, the ambient intensity, tension, and attentive focus to particular musical textures would also disappear—and the phenomenal character of the music-as-given to our lone listener would be dramatically altered. Try as we might, we simply cannot recreate this atmosphere within our solitary listening episodes. The experiential character of deep listening is thus dramatically different when others are involved.

To continue with this example: an integral part of many live music experiences is dancing. Dancing is a robustly embodied response to musical events. Moreover, it is the enactment of an attentional focusing that shapes how and what we hear. Via dancing, the temporal regularities of melodic and rhythmical patterns within the music are physicalized within an array of bodily movements. And the coordination between sonic pattern and bodily movement—an instance of bodily entrainment—is an enactive gesture, a perceptual exploration of the piece's sonic topography. Again, bodily movements modulate the listener's relation to different features of the piece (e.g. metre and melody); dancing experientially foregrounds these features and shapes the way that these features stand out against the background of the piece's other sound features. The temporal predictability and consistency of sonic patterns afford this sort of bodily engagement. Sonic patterns therefore afford an entering into the inner recesses of sonic space, a point of access for losing ourselves, experientially, within the piece via the immersive 'deep listening' that often occurs whilst dancing.

Yet dancing at a live music performance is not simply an instance of the listener being aware of and responding to solitary possibilities for musical interaction. Additionally, part of the content of the dancer's awareness is the dancing-responses of *other* dancers. Their attunement and reaction to sonic patterns shapes the listener's *own* experience of these patterns. Dancing is thus a vehicle of joint attention, a means of enacting a shared attentional framework that shapes the character and content of the piece-as-perceived in that context.[9] When others' dancing reactions shift, for instance, I feel my body compelled to alter my own movements accordingly. I bob my head in time and sway my body, carried along by the pulse and tempo of the crowd's movement. In this way do I come to inhabit the lived time and structural space of the musical piece with others. The rhythmical and sonic patterns of a piece, and the dancing these patterns afford, forge an interactive phenomenon that synchronizes the joint listeners to one moving mass (Vickhoff and Malmgren 2004, p. 19). And this one moving mass enacts a shared attentional framework unique to that time, place, and performance. However, both the individual as well as collective integrity of the experience simultaneously coexist within that musical experi-

---

9   See Cochrane (2009) for more on joint attention and shared musical experience.

ence. Participants within that shared experience are able to interpret a steady flow of musical features and patterns in individual terms, while the temporal regularities of the sonic invariants discussed coordinate their individual behaviour as well as their attentional foci (Cross 2006, p. 122).

This enactive characterization of musical experiences, both solitary and shared, emphasizes the central role that agency plays in shaping both the character and content of musical experience — including its spatial content. Having offered some phenomenological descriptions of different forms of deep listening, I now want to look at supplementary empirical evidence that seems to support these descriptions.

5 Empirical support

The first line of empirical evidence I want to look at concerns amusia. Amusia is profound tone deafness, an inability to hear music as music. More formally, it is a severe deficiency in processing pitch variation despite normal speech perception and intact sense of rhythm (Ayotte *et al.* 2002; Peretz *et al.* 2002; Sacks 2007). There are different forms of amusia. For the total amusiac, however, music is experienced as incoherent noise, an irritating sound structure lacking any sort of aesthetically-compelling character. For example, one amusiac described Rachmaninov's second piano concerto as sounding like 'banging and noise' (McDonald and Stewart 2008), whereas another describes the experience of listening to music as akin to a screeching car (Sacks 2007, p. 101).

The conventional explanation of amusia portrays it as an auditory deficit (1) related to deficiencies in fine-grained processing of musical pitch variations, and (2) confined to the musical domain and musical abilities (Ayotte *et al.* 2002). However, some recent studies challenge this characterization. They propose instead that amusia is not a specifically sensory-musical deficit but rather a *spatial* deficit — that is, an inability to represent space (Cupchik *et al.* 2001; Douglas and Bilkey 2007; Särkämö *et al.* 2009). For instance, amusiacs were found to perform significantly worse than non-amusiac controls on mental rotation tasks (Douglas and Bilkey 2007). Cupchik *et al.* (2001) found a correlation between performance on a mental rotation task involving three dimensional figures and the ability of the listener to perceive inverse and retrograde musical permutation (i.e. when a musical piece had been played backwards). Whether or not amusia stems from a spatial deficit is a matter of some debate (see, e.g. Tillmann *et al.* 2010). However, if something like this is the case, it lends insight into the amusiac's musical phenomenology — or rather, lack thereof. For, it seems that amusiacs are unable to perceive music as offering up the spatially-inviting auditory profile that normal listeners perceive. They might perceive the outer or *locational* spatial profile of music specifying its egocentric location (that 'banging and noise' is coming from *over there*). But they are unable to perceive a piece's inner or *structural* spatial profile—that is, the spatial quality that specifies that sound event's uniquely musical character. And without the ability to enact the spatial fusing involved in deep listening, music thus remains an alien and impenetrable entity.

This view receives support from another study. As discussed above, the animate body plays a central role in the spatial fusing characteristic of deep listening. Again, bodily movements such as swaying back and forth, nodding our heads, tapping fingers and toes, or the more energetic whole-body dynamics of dancing, modulate our perception of the spatial dimensions of musical experience. In particular, bodily synchronization with rhythmic patterns and tempo—actions that, as we'll see in a moment, we seem born ready and able to enact—open up the inner space of a piece. This bodily engagement with music is both a response to and an affirmation of music's spatially-structured exploratory profile. Amusiacs, however, have a marked difficulty in synchronizing bodily movements with music—despite a normal ability to synchronize with sequences of non-musical sounds (Dalla Bella and Peretz 2003). Another more recent study affirmed this result, indicating that the deficit in processing rhythmic patterns was related not to the complexity of the patterns themselves—the subjects were able to synchronize with monotonic sounds such as a steady drum beat, for example—but rather the pitch-variations of the music (Foxton *et al.* 2006). Again, lacking the ability to perceive and respond to the inner structural space of the musical piece, the amusiacs were accordingly unable to enact a robust sensorimotor response to the music—which in turn affected both the experiential character of the music-as-perceived (i.e. as having a disagreeable sonic character) as well as the music-as-experiential content (i.e. as an impenetrable object lacking a spatially-inviting exploratory profile).

Daniel Vastfjall (2003) found that both the experienced presence of music (i.e. sound immersion, or the feeling of involvement with a piece) as well as experienced emotions in response to music varies as a function of its perceived spatiality. In other words, the spatial content of musical experience, in contrast to other acoustic parameters (e.g. pitch, timbre, loudness, etc.), is arguably what triggers the profound immersive and emotional responses characteristic of deep listening. In Vastfjall's study, participants were seated and asked to close their eyes and listen in a concentrated (i.e. 'deep') way—they were told to 'let themselves into the music'—focusing in particular on the intensity of their emotional reactions to the pieces (Vastfjall 2003, p. 184). Predictably, participants in the 'mono condition' (i.e. listening to music via one-channel) did not respond to emotion induction to the same extent as participants in either the stereo (i.e. two-channel, or speakers on either side) or six loudspeakers conditions (*ibid.*, p. 185). In the latter conditions, the music was experienced as more immersive and thus more emotionally compelling. It appears, then, that the 'subjective sense of presence and emotional reactions to the music are highly interrelated' — affirmed by the fact that '[p]articipants who experienced a strong feeling of presence and a sense of being in the sound field also reported stronger emotional reactions' (*ibid.*, p. 186). When the spatial content of musical experience is absent or diminished — such as with amusia, or perhaps in more common inattentive or shallow modes of listening — the immersive and emotional character is also compromised. Likewise, Don Ihde quotes a philosopher friend who recalls first becoming aware of his increasing deafness when he began to lose interest in music, which gradually became 'distant... objectlike... over there apart from me' (Ihde 2007, p. 78). In this case, music appears to have lost its spatial character; what was once a dynamic, spatially-structured soundworld was reduced to an inert acoustical object. Deep listening was no longer possible, only observational hearing.

With practice and experience, one can presumably cultivate and refine the attentional and sensorimotor skills needed for deep listening. However, multiple streams of empirical evidence from neonate music therapy suggest that we are potentially deep listeners from birth. This is not the place for a comprehensive review of the literature (see, e.g. Standley 2001; for discussion, see Krueger forthcoming a,b). But we can note a few salient points. Generally speaking, music therapy consists of a cluster of various music-related practices and techniques designed to give patients of all ages the opportunity to explore and communicate emotions (Bunt and Pavlicevic 2001, p. 181). Traditionally geared toward children and adults with various disabilities or mental health problems, the past few decades have seen a rising interest in music's therapeutic effect on neonates. Specifically, neonate music therapy has arisen in response to what Tia DeNora terms the 'paradox of the NICU' (DeNora 2000). This is the idea that the hostile soundworld of the NICU — comprised of, for example, the auditory byproduct of medical technologies (e.g. respirators, bottles clanking on incubators, noisy beeps of heart monitors and other machinery amplifying aspects of the infant's disorganized state, etc.), the sound of other infants crying, the continual commotion of people moving in and out of the area — might actually be disrupting the infant's basic life-processes, in turn affecting sleep regularization and state lability (*ibid.*, p. 80; see also Haslbeck 2004; Kaminski and Hall 1996, p. 46). However, a significant amount of research seems to indicate that music can be a valuable resource for enhancing the neonate's physiological and emotional well-being, serving as an occluding corrective to this unfriendly sound environment (Standley 2001, p. 213; DeNora 2000, p. 81).

For the purposes of this discussion, I am most interested in how neonates and infants enact musical experience — that is, how infants seem to perceive music as presenting a spatially-structured exploratory profile inviting bodily entrainment. From the start, infants are surprisingly skilled listeners, seemingly attuned to the rhythmic, emotional, and communicative opportunities that musical engagement offers.[10] Like adults, they appear to appreciate and respond to music as an experientially salient feature of their perceptual environment. For example, both term and pre-term infants attend more fixedly to music than they do to other ambient noises, suggesting a preference for the sonic coherence and organizational structure of music in contrast to contingent environmental noise (Butterfield and Siperstein 1972; Standley 2001). Infant activity tends to decrease in response to auditory stimuli generally. But the most significant decreases are caused by music, further suggesting that music is a preferred auditory stimulus (Kagan and Lewis 1965). Other studies have found that infants are surprisingly discriminating listeners. Not only do infants tune in to overarching musical patterns, preferring consonant over dissonant intervals (Trainor and Heinmeiller 1998, p. 83; Zentner and Kagan 1998). Additionally, they are able to pick out and attend to fine-grained auditory properties of music such as pitch, melody, tempo, and musical phrase structure (Schellenberg and Trehub 1996; Trehub *et al.* 1999; Trehub and Schellenberg 1995; Trehub and Trainor 1993). For example, three- to six-month-olds can vocalize a matched pitch to sung tones (Wendrich 1981) as well as learn to turn toward a loudspeaker whenever they perceive a change in background melody (Trehub *et al.* 1987). Two-month-olds can remember short melodies and discriminate it from other melodies (Plantinga and Trainor 2009). Infants, it would seem, are therefore capable of hearing and responding to the particular sound features that carry a piece's expressive content. Even the very young possess the perceptual skills needed to find music perceptually captivating because of its emotional expressivity (Nawrot 2003). Additionally, they possess the listening skills needed to actively *explore* music, to *enact* musical experience, by selectively attending to and bodily engaging with aspects of its sonic topography.

To further see how this is so, consider that, beyond merely exhibiting the perceptual skills needed to make musical discriminations, infants also seem to experience music as affording communicative possibilities. This is indicated by their rhythmic bodily entrainment responses to music. Haslbeck (2004), for example, found that, over the course of several music therapy sessions, the pre-term neonates in her study gradually became active participants within the sessions, intentionally seeking interpersonal contact via the music, which consisted of slowly-sung melodies supplemented with a hand resting gently on the infant's chest or back. This inter-

---

10 Again, considerations of space prohibit an adequate defence of this thesis here. See instead Krueger (forthcoming b).

personal contact emerged via bodily entrainment: the infants enacted whole-body 'rhythmic dialogues' with the music (*ibid.*, p. 9). These dialogues were established via a coordinated rhythmic alteration between the sung lullaby and the infants' bodily responses. For instance, both sucking/swallowing and regularized patterns of respiration were observed to mimic the rhythmic alterations of the sung melody (e.g. sucking at the end of melodic phrases—*ibid.*, p. 9). More tellingly, the infants gradually initiated eye contact, summoned an increasingly attentive and engaged posture (while reducing fidgeting and grimacing), and exhibited increased mouth movements (playing with the tongue, mouthing the vowels being sung, such as 'o' and 'u') and vocalizations during the sessions. Other movements included opening and closing of hands, wrinkled brows, and eyes opening and closing in sync with the rising of falling of the sung melody (*ibid.*, p. 11). This opportunity for social contact within music therapeutic contexts is crucial—and, seemingly, something that the infants in Haslbeck's sessions actively sought out—given the affective isolation of the pre-term infant's life inside the incubator. Additionally, it suggests that infants are enactively attuned to the spatial characteristics of music since, without this attunement—as the amusia research demonstrated—this sort of rhythmic entrainment cannot occur.[11]

What the evidence discussed above indicates, I suggest, is that particular structural features of music—again, textural qualities and regularities of melodic and rhythmic patterns—are actively perceived and exploited by infants in episodes of proto-deep listening.[12] This is further confirmed by research indicating that movement influences auditory encoding of rhythm patterns in both infants and adults. How we move shapes both what we hear and how we hear it. In a series of experiments, Jessica Philips-Silver and Laurel Trainor trained 7-month-old infants by listening to an ambiguous two-minute rhythmic pattern (i.e. a pattern lacking accented beats). Half of the infants were bounced on every second beat and half were bounced on every third beat. As a result, the infants expressed a more prolonged interest in the auditory test stimulus with the metrical form—every second beat accented (the duple form) in one stimulus, and every third beat (the triple form) in the other—that matched the metrical form of their training bouncing (Philips-Silver and Trainor 2007, p. 1430). This was also the case when blindfolded. A further experiment

11 Another study found that preverbal infants spontaneously display tempo-sensitive rhythmic motions of their body with music—and that this bodily engagement is a source of positive affect—but that they don't exhibit this behaviour in response to speech or other arrhythmic ambient noise (Zentner and Eerola 2010). There are also indications that some animals (e.g. parrots) are capable of similar coordination (Patel *et al.* 2009; Schachner *et al.* 2009). Perhaps they, too, are capable of a kind of proto-deep listening.

12 Gentle music with 'thin' textures (e.g. sung lullabies, new age music, etc.) and no abrupt modulations of volume or tempo reduces alerting responses in infants, offering up a stable, inviting, and predictable soundworld for the infant to explore (Standley 2001; 2002).

showed that personal bodily movement was necessary to establish this metrical preference. Watching the experimenter bounce during the ambiguous rhythm training failed to establish a preference for either of the auditory stimulus versions (*ibid.*, p. 1430).[13] A similar set of experiments was later done with adults (*ibid.*). Unlike the infants, of course, the adults could engage in their own 'bounce training'. But like the infants, the adults' synchronized movements of their body determined how they heard an ambiguous musical rhythm (*ibid.*, p. 543). Once again, they had to personally bounce their own bodies, and not watch a video of another doing it, in order for their experience of the ambiguous rhythm to covary relative to their particular bounce training (e.g. bouncing on every second or on every third beat). But their sensorimotor training determined how they enacted the content of their experience of the ambiguous rhythm. Ample empirical evidence therefore suggests that even infants possess rudimentary (i.e. practical or pre-theoretical) sensorimotor understanding of how modulations of bodily movement and attentional focusing affect sensory change. They are capable of enacting rich musical experiences from the start. Moreover, music is perceived, again from the start, in terms of its spatial character. For the normal listener, music manifests experientially as harbouring non-derived spatial content.

6 An objection

I now want to briefly consider a natural objection to the enactive view defended above. We can term this the 'immobile listener objection'. This is the objection that listeners with various sorts of extreme sensorimotor deficits (e.g. quadriplegics, individuals with Locked-in Syndrome, etc.) lack the ability to enact a robust sensorimotor engagement with music. Yet they nevertheless clearly perceive music as music—and surely, moreover, are capable of being moved deeply by it, experiencing it in an immersive and emotionally resonant way. Thus, music listening cannot depend essentially upon exploratory sensorimotor skills and actions the way this chapter has argued that is does.

However, this objection misses the mark for a couple of reasons. First, many sufferers of spinal cord injuries or paralysis had extensive periods of perceiving prior to suffering their injury (e.g. as the result of a fall or car accident) or the onset of, for example, Multiple Sclerosis in young adulthood or Locked-in Syndrome later in life. So, they clearly retain a practical understanding of how movement and attentional focusing modulates sensory change—even once their movement is inhibited. Moreover, they retain a range of practical skills needed to enact experience of different sorts, such as the ability to move their eyes, head, and (with the assistance

---

13   For samples of the experimental sound stimuli, see the following link: http://www.sciencemag.org/cgi/content/full/308/5727/1430

of technology such as a wheelchair) their entire bodies in relation to their environment. Despite their physical limitations, quadriplegics lead active exploratory lives. They are continually 'engaged in the task of orienting themselves in relation to the world around them and to gravity' (Noë 2004; see also Cole 2004). Thus, they remain active perceivers even if the range of their active perceiving is somewhat restricted. And in the case of music perception, those in wheelchairs are entirely capable of enacting rhythmic synchronization with music — which, as we've seen, is a crucial enactive gesture for opening up the spatial character of the auditory event *qua* music, a process which affords deeper, more focalized listening. Not only can they nod their heads or blink in time with music, or attentionally follow the contour of a melody as it moves through tonal space. Additionally, wheelchairs can be summoned to perform all sorts of skilled, active engagements with music — swaying back and forth, twirling in circles, tilting from side to side, etc. — as an internet video search will quickly reveal. These movements allow the wheelchair-bound listener to explore how the dynamics of embodied engagement alter the character and content of musical experience. Wheelchair-bound perceivers are thus far from immobile listeners. To the contrary, they remain capable of enacting rich spatially-structured musical content. They retain the skills and practical understanding needed to respond to the unique exploratory profile music offers; they have the skills to access music in a sensitive or 'deep' manner. It is rather those with a *spatial* deficit (e.g. amusiacs) who can no longer perceive and respond to this exploratory profile and thus who have lost this ability. In the latter cases, both the character and content of musical experience has been dramatically altered, as the perceptual reports of amusiacs would seem to indicate.

7 Concluding thoughts

I have argued that listening to sounds is an active sensorimotor exploration of our world — including spatial and locational aspects of that world and things in it. Sounds routinely furnish spatial information about our world, and we use our auditory experiences to explore and skillfully respond to this information. This is particularly evident in the case of music perception. Specifically, I have argued that we enact our musical experience — that is, we summon a range of bodily skills to secure experiential access to music, and, in particular, its spatially-complex character. For it is here that the source of music's experiential richness lies. This complex spatial character — as well as the way that this complex spatial character is articulated in and through music's temporal dynamics — is what makes music such a uniquely compelling phenomenon. I have tried to bring this out with a phenomenological characterization of what I termed 'deep listening', focusing in particular on how bodily skills and active perceptual exploration play a central role in enacting musical content. The account developed above is, of course, merely a sketch. But for enactive accounts of perception to develop, they must extend the discussion beyond the well-tread terrain of vision and touch and move into the domain of other sensory modalities. Considering the nature of music perception, I suggest, is a particularly fertile way to do just this.

References

Ayotte, J., I. Peretz and K. Hyde (2002), 'Congenital Amusia: A Group Study of Adults Afflicted with a Music-Specific Disorder', *Brain*, **125**: 238–51.

Becker, J. (2004), *Deep Listeners: Music, Emotion, and Trancing*, Bloomington (IA), Indiana University Press.

Bunt, L. and M. Pavlicevic (2001), 'Music and Emotion: Perspectives from Music Therapy' in P.N. Juslin and J.A. Sloboda, Eds., *Music and Emotion: Theory and Research*, Oxford, Oxford University Press.

Butterfield, E. and G. Siperstein (1972), 'Influence of Contingent Auditory Stimuli Upon Nonnutritional Suckle' in J. Bosma, Ed., *Oral Sensation and Perception: The Mouth of the Infant*, Springfield (IL), Charles C. Thomas: 313–33.

Carr, C. (2002), 'Neuroscience: Sounds, Signals and Space Maps', *Nature*, **415** (6867): 29–31.

Casati, R. and J. Dokic (2009), 'Some Varieties of Spatial Hearing' in M. Nudds and C. O'Callaghan, Eds., *Sounds and Perception: New Philosophical Essays*, Oxford, Oxford University Press: 97–110.

Cochrane, T. (2009), 'Joint Attention to Music', *British Journal of Aesthetics*, **49** (1): 59–73.

Cole, J. (2004), *Still Lives: Narratives of Spinal Cord Injury*, Cambridge (MA), MIT Press.

Cross, I. (2006), 'Music and Social Being', *Musicology Australia*, **XXVIII**: 114–26.

Cupchik, G.C., K. Phillips and D.S. Hill (2001), 'Shared Processes in Spatial Rotation and Musical Permutation', *Brain and Cognition*, **46** (3): 373–82.

Dalla Bella, S. and I. Peretz (2003), 'Congenital Amusia Interferes with the Ability to Synchronize with Music', *Annals New York Academy of Sciences*, **999**: 166–9.

DeNora, T. (2000), *Music in Everyday Life*, Cambridge, Cambridge University Press.

Douglas, K.M. and D.K. Bilkey (2007), 'Amusia is Associated with Deficits in Spatial Processing', *Nature Neuroscience*, **10** (7): 915–21.

Foxton, J.M., R.K. Nandy and T.D. Griffiths (2006), 'Rhythm Deficits in "Tone Deafness"', *Brain and Cognition*, **62** (1): 24–9.

Gangopadhyay, N. and J. Kiverstein (2009), 'Enactivism and the Unity of Perception and Action', *Topoi*, **28** (1): 63–73.

Hamilton, A. (2009), 'The Sound of Music' in M. Nudds and C. O'Callaghan, Eds., *Sounds and Perception: New Philosophical Essays*, Oxford, Oxford University Press: 146–82.

Haslbeck, F. (2004), 'Music Therapy with Preterm Infants—Theoretical Approach and First Practical Experience', *Music Therapy Today (online)*, **5** (1): 1–15.

Hurley, S. (2001), 'Perception and Action: Alternative Views', *Synthese*, **129** (1): 3–40.

Hyde, P.S. and E.I. Knudsen (2002), 'The Optic Tectum Controls Visually Guided Adaptive Plasticity in the Owl's Auditory Space Map', *Nature*, **415** (6867): 73–6.

Ihde, D. (2007), *Listening and Voice: Phenomenologies of Sound*, Albany (NY), SUNY Press.

Kagan, J. and M. Lewis (1965), 'Studies of Attention in the Human Infant', *Journal of Developmental Psychology*, **11** (2): 95–127.

Kaminski, J. and W. Hall (1996), 'The Effect of Soothing Music on Neonatal Behavioral States in the Hospital Newborn Nursery', *Neonatal Network*, **16**: 45–54.

Kierkegaard, S. (1959), *Either/Or* (Vol. 1), D.F. Swenson, Trans., Garden City (NY), Double Day.

Krueger, J. (forthcoming a), 'Doing Things with Music', *Phenomenology and the Cognitive Sciences*.

Krueger, J. (forthcoming b), 'Empathy, Enaction, and Shared Musical Experience' in T. Cochrane, B. Fantini and K.R. Scherer, Eds., *The Emotional Power of Music:*

Multidisciplinary Perspectives on Musical Expression, Arousal and Social Control, Oxford, Oxford University Press.

McDonald, C. and L. Stewart (2008), 'Uses and Functions of Music in Congenital Amusia', *Music Perception*, **25** (4): 345–55.

Merleau-Ponty, M. (2002), *Phenomenology of Perception*, C. Smith, Trans., New York, Routledge.

Merleau-Ponty, M. (2004), *The World of Perception*, New York, Routledge.

Morgan, R.P. (1980), 'Musical Time/Musical Space', *Critical Inquiry*, **6** (3): 527–38.

Nawrot, E.S. (2003), 'The Perception of Emotional Expression in Music: Evidence from Infants, Children and Adults', *Psychology of Music*, **31** (1): 75–92.

Noë, A. (2004), *Action in Perception*, Cambridge (MA), MIT Press.

Noë, A. (2009), 'Conscious Reference', *The Philosophical Quarterly*, **59** (236): 470–82.

Nonken, M. (2008), 'What Do Musical Chairs Afford? On Clarke's Ways of Listening and Sacks's Musicophilia', *Ecological Psychology*, **20** (4): 283–95.

Nudds, M. (2001), 'Experiencing the Production of Sounds', *European Journal of Philosophy*, **9** (2): 210–29.

O'Callaghan, C. (2009), 'Auditory Perception' in E.N. Zalta, Ed., *The Stanford Encyclopedia of Philosophy*, [Online], http://plato.stanford.edu/archives/sum2009/entries/perception-auditory/

O'Callaghan, C. (2010), 'Perceiving the Locations of Sounds', *Review of Philosophy and Psychology*, **1** (1): 123–40.

O'Regan, J.K. and A. Noë (2001), 'A Sensorimotor Account of Vision and Visual Consciousness', *Behavioral and Brain Sciences*, **24** (5): 939–73.

O'Shaughnessy, B. (1984), 'Seeing the Light', *Proceedings of the Aristotelian Society*, New Series, **85**: 193–218.

O'Shaughnessy, B. (2009), 'The Location of a Perceived Sound' in M. Nudds and C. O'Callaghan, Eds., *Sounds and Perception: New Philosophical Essays*, Oxford, Oxford University Press: 111–25.

Patel, A.D., J.R. Iversen, M.R. Bregman and I. Schulz (2009), 'Experimental Evidence for Synchronization to a Musical Beat in a Nonhuman Animal', *Current Biology*, **19** (10): 827–30.

Peretz, I., J. Ayotte, R.J. Zatorre, J. Mehler, P. Ahad, V.B. Penhune and B. Jutras (2002), 'Congenital Amusia: A Disorder of Fine-Grained Pitch Discrimination', *Neuron*, **33** (2): 185–91.

Plantinga, J. and L.J. Trainor (2009), 'Melody Recognition by Two-Month-Old Infants', *The Journal of the Acoustical Society of America*, **125** (2): 58–62.

Philips-Silver, J. and L.J. Trainor (2007), 'Hearing What the Body Feels: Auditory Encoding of Rhythmic Movement', *Cognition*, **105** (3): 533–46.

Putnam, H. (1999), *The Threefold Cord: Mind, Body, and World*, Columbia (NY), Columbia University Press.

Reybrouck, M. (2005), 'A Biosemiotic and Ecological Approach to Music Cognition: Event Perception Between Auditory Listening and Cognitive Economy', *Axiomathes*, **15**: 229–66.

Sacks, O. (2007), *Musicophilia: Tales of Music and the Brain*, London, Picador.

Särkämö, T., M. Tervaniemi, S. Soinila, T. Autti, H.M. Silvennoinen, M. Laine and M. Hietanen (2009), 'Cognitive Deficits Associated with Acquired Amusia After Stroke: A Neuropsychological Follow-Up Study', *Neuropsychologia*, **47** (12): 2642–51.

Schachner, A., T.F. Brady, I.M. Pepperberg and M.D. Hauser (2009), 'Spontaneous Motor Entrainment to Music in Multiple Vocal Mimicking Species', *Current Biology*, **19** (10): 831–6.

Schellenberg, E.G. and S.E. Trehub (1996), 'Natural Musical Intervals', *Psychological Science*, **7** (5): 272–7.

Schopenhauer, A. (1966), *World as Will and Representation*, E. Payne, Trans., (Vol. 1), Mineola (NY), Dover Publications.

Sloboda, J.A. (2005), *Exploring the Musical Mind: Cognition, Emotion, Ability, Function*, Oxford, Oxford University Press.

Small, C. (1998), *Musicking*, Middletown (CT), Wesleyan University Press.

Smith, F.J. (1979), *The Experiencing of Musical Sound: Prelude to a Phenomenology of Music*, New York, Gordon and Breach.

Standley, J. (2001), 'Music Therapy for the Neonate', *Newborn Infant Nursing Reviews*, **1** (4): 211–6.

Standley, J.M. (2002), 'A Meta-Analysis of the Efficacy of Music Therapy for Premature Infants', *Journal of Pediatric Nursing*, **17** (2): 107–13.

Thompson, E. (2005), 'Sensorimotor Subjectivity and the Enactive Approach to Experience', *Phenomenology and the Cognitive Sciences*, **4** (4): 407–27.

Tillmann, B., P. Jolicœur, M. Ishihara, N. Gosselin, O. Bertrand, Y. Rossetti and I. Peretz (2010), 'The Amusic Brain: Lost in Music, but Not in Space', *PLoS ONE*, **5** (4): e10173.

Trainor, L.J. and B.M. Heinmiller (1998), 'The Development of Evaluative Responses to Music: Infants Prefer to Listen to Consonance Over Dissonance', *Infant Behavior and Development*, **21** (1): 77–88.

Trehub, S.E. and E.G. Schellenberg (1995), 'Music: Its Relevance to Infants', *Annals of Child Development*, **11**: 1–24.

Trehub, S.E., E.G. Schellenberg and S. Kamenetsky (1999), 'Infants' and Adults' Perception of Scale Structure', *Journal of Experimental Psychology*, **25** (4): 965.

Trehub, S.E., L.A. Thorpe and B.A. Morrongiello (1987), 'Organizational Processes in Infants' Perception of Auditory Patterns', *Child Development*, **58** (3): 741–9.

Trehub, S.E. and L. Trainor (1993), 'Listening Strategies in Infancy: The Roots of Music and Language Development' in S. McAdams and E. Bigand, Eds., *Thinking in Sound: The Cognitive Psychology of Human Audition*, Oxford, Oxford University Press.

Turner, P., S. Turner and I. Mcgregor (2007), 'Listening, Corporeality and Presence', presented at the *Presence 2007: 10th Annual Workshop on Presence*, Barcelona, Spain, [Online], http://citeseerx.ist.psu.edu/viewdoc/summary?doi=10.1.1.92.7329

Vastfjall, D. (2003), 'The Subjective Sense of Presence, Emotion Recognition, and Experienced Emotions in Auditory Virtual Environments', *CyberPsychology & Behavior*, **6** (2): 181–8.

Vickhoff, B. and H. Malmgren (2004), 'Why Does Music Move us?', *Philosophical Communications, Web Series*, **34**, Dep. of Philosophy, Goteborg University, Sweden.

Wendrich, K. (1981), *Pitch Imitation in Infancy and Early Childhood: Observations and Implications*, Storrs (CT), University of Connecticut.

Zahorik, P. and F.L. Wightman (2001), 'Loudness Constancy with Varying Sound Source Distance', *Nature Neuroscience*, **4** (1): 78–83.

Zentner, M.R. and T. Eerola (2010), 'Rhythmic Engagement with Music in Infancy', *Proceedings of the National Academy of Sciences of the United States of America*, [Online], http://www.pnas.org/content/early/2010/03/08/1000121107.full.pdf+html?frame=header

Zentner, M.R. and J. Kagan (1998), 'Infants' Perception of Cosonance and Dissonance in Music', *Infant Behavior and Development*, **21** (3): 483–92.

Liliana Albertazzi

Extended Space in Perception and Art

1 Internalism and externalism at the crossroads

In my view, the contemporary debate in mind theory and more generally cognitive science on externalism (see Clark 1998; 2008; Davidson 1980; Gallagher 2009; O'Reagan and Nöe 2001; Honderich 2006; Manzotti 2009; among others) and internalism (Fodor 1983; Searle 1980; 1983) has structural weaknesses: firstly because it is confined to the specific original context and risks becoming self-referential, and secondly because it makes a categorial mistake.

The internalism in question grew out of the science of AI; it represents the mind as an internal mechanism (computer) for the elaboration, transformation and representation of stimuli (metrical cues) originating in transcendent reality (sometimes identified with classical mechanics). The development of neuroscience essentially changed the reference from one kind of computer to another (the brain), which acts primarily according to inferential-probabilistic principles rather than logical-deductive ones. Aside from the specific differences in research and instruments of investigation, in both cases the internalism of the mind refers to a computational mechanism and is analysed primarily according to computational methods.

An interesting internalist proposal that grew out of neuroscience is the embodied constructivism developed by Maturana and Varela, who presented a conception close to the main tenets of phenomenology (Maturana and Varela 1980) while underlining the shallowness of a representationalist theory of reality (usually advocated by the promoters of symbolic internalism, Fodor 1983; Chomsky 1968; Piattelli-Palmarini 1980; Pylyshyn 1981a,b). Other proposals in favour of the primacy of experience over stimuli have been made, in line with Merleau-Ponty's phenomenology (Merleau-Ponty 1962; 1968): for instance, Gibson's 'direct' perception (Gibson 1950; 1979) and a more recent proposal of an embodied approach to perception (O'Regan and Noë 2001; Noë 2004; 2009; Thompson 2005), which upholds a theory of perception as action while at the same time sustaining the idea of an extended mind (see also Clark and Chalmers 1998).

It can be seen that the term 'embodied' is not used unequivocally in the various proposals. Rather, the use of language tends to group together different theories that do not always share the same epistemological or even ontological hypotheses on the relationship connecting cognitive processes and structures to reality, both at the primary (i.e. perceptual) and secondary (i.e. mental) level (for this distinction see Kanizsa 1991). Several quite distinct conceptions proclaim themselves in favour of an embodied or situated consciousness, such as inverse optics, the inferential theory of perception, naïve realism of the Gibsonian kind, and the Gestaltist conception. Not all of these share the same idea of reality or even the same idea of representation, which ranges from internalist hypotheses (the aforementioned enactive perception as autopoiesis of Maturana and Varela's system) to externalist hypotheses. Lastly, nearly all the versions of embodied cognition, whether internalist or externalist, reduce the mind to the brain and/or to a psychophysical body and hence share a fundamental reductionism that uses classical physics or biology as its primary ontological referent for the explanation of phenomena. Because they lack a precise distinction between the various levels of reality (Poli 2001; 2006), these theories tackle the issue of mind and consciousness in quantitative terms, i.e. in terms of stimuli and of the elaboration of information contained within the stimuli according to the classic conception of Shannon and Weaver (1949/1998). Embodied and enactive approaches can be powerful methodological tools in behavioural and neurophysiological investigation, but their ultimate reliance on sensorimotor contingencies for the construction of mental content inevitably gives them a naïve realist flavour (see Vishwanath 2005. More on this topic in Albertazzi, van Tonder and Vishwanath 2011, § 4).

Other proposals are instead still couched in terms of inferences or symbolic-conceptual interpretations of stimuli (Rock 1983; 1999; Gregory 1997), which avoid the fact and the explanation of an evident and direct perception of meaningful appearances in daily life.

In all theories, internalist or externalist, representationalist or embodied, past experience plays a prime role in the recognition of objects as an unconscious process or in terms of instances of the corresponding concept, based on characterizations or interpretations of a linguistic/cognitive nature, be they mental images (Johnson-Laird 1983; Jackendoff 1992) or pictures in the sense of computational vision theory (Marr 1982), i.e. digital images. Linguistic concepts and categories are considered to not only reinforce perception, as for instance in the case of the categories of colour (Agrillo and Roberson 2007; Boroditsky 2001; Davidoff 1991; 2006; Kay and Kempton 1984), but even to constitute the meaning, i.e. the way in which *the processed stimuli are interpreted*. The difference between the various theories lies in the weight placed on the symbolic representational models of the *brain* (Kosslyn 1980; 1994; Kosslyn and Pomerantz 1977). The boundaries between mind and brain remain ill defined and can only be traced in light of the previous distinction, i.e. a preliminary separation of the field of analysis.

To date, all approaches in mind and perception theory, regardless of their theoretical premises and of which particular flavour of internalism or externalism to which they refer, have begun their enquiries from the same metric primitives (cues) that, for better or worse, have characterized the Galilean approach to science — an approach that has been so far extremely successful because it has enabled identification of the rules obeyed by several physical and biological phenomena, leading to extraordinary capabilities in their manipulation and prediction through the staggering development of new techniques. What has so far remained completely beyond these bounds is the science of qualitative aspects of experience (Pinna 2008), including aesthetics, which remain sundered in principle and in fact from the primary qualities on which Galilean science is based.

2 Beyond the Galilean approach

Already in his time, however, Helmholtz observed that privileging the metric or primary qualities as primitives of perception gave rise to 'astonishing consequences'. For example the fact that:

> the objects present in space appear to us 'clothed' in the qualities of our sensations. They appear to us red or green, cold or warm, they have smell or taste, etc., while these sensory qualities belong, after all, only to our nervous system and do not extend into outer space. (Helmholtz 1867)

In short, if it is assumed that the information that serves to produce the perceived object derives from the stimuli and that the percepts are an *internal* product of our nervous system, one fails to understand how the space 'outside' of us or 'in front of us' is generated, or why there is a necessary *externalization* of the objects of vision. At issue is precisely the Galilean paradigm, and its conception of space as unitary, that is, non-differentiated between experiential/perceptual space and physical space (Albertazzi 2006b). Assuming a Galilean paradigm in science, in fact, has heavy consequences. As Köhler observed:

> the physicist of the late baroque period [i.e. the Galilean] *subtracts many so-called sensory qualities* (secondary qualities) if he wants to extract what he considers the objective realities from the phenomenal manifolds... but under no circumstances has the *phenomenal object* anything to do with the place in physical space where the '*corresponding*' physical object is located... also its localization in the brain is ruled

out, because, without any doubt, we have *the phenomenal object before us and outside of ourselves*. (Köhler 1929/1971)

The question is still entirely unresolved, and it is also present in more recent areas of enquiry like neuroscience. As Crick and Koch noted, for example, what the brain apprehends in our perception usually concerns the external world or other body parts. This is why what we see seems to be localized *outside* ourselves, despite the fact that the neurons 'which see' (according to Crick and Koch's approach) are to be *internally* localized *in our head* (Crick and Koch 1998). In this conception, the external world and space are identified with physical reality, to which the mind — or, more accurately, neuronal activity — refers.

Further areas of research, such as virtual reality, present the same problem, albeit in a meta-world of simulation and representation (Sobchack 2004). In fact, at the core of the successful development of 3D technology lies an understanding of the visual processing that underpins our ability to perceive our 3D world. The ultimate goal of such technology, especially virtual reality, is to make oneself feel that one is 'in a real scene'; in other words, we want to make virtual reality as phenomenal as phenomenal experience: the problem is that we do not 'know' what phenomenal experience and qualities are, in the sense that there is not (yet) a science of them. For the time being, however, what we know is the psychophysical correlation between stimuli and behavioural response, between stimuli and (some) neural correlates. It is evident, however, that there is much more to phenomenological perception than the stimulus understood in the neurophysiological sense.

This problem in perception and cognition studies also has a bearing on other qualitative analyses such as aesthetics, which has recently been given renewed momentum by so-called neuroaesthetics (Zeki 1999; see also Livingstone 2002; Solso 2003; Ramachandran 2003) and by cognitive and psychophysical analyses (McManus, Cook and Hunt 2010). While trying to formulate an experimental aesthetics, these disciplines, too, have attempted an analysis and an explanation of the aesthetic — and thus qualitative — meaning of artworks based on the activity of psychophysical or neuronal correlates.

Experimental aesthetics was of course not born yesterday, for it dates back to the first days of classical psychophysics (Fechner 1876). Modern aesthetics shares a categoric error with Fechner's experimental aesthetics (Albertazzi 1997; Massironi 2000) in that it attempts to explain the nature of complex phenomena of a qualitative nature based on phenomena of a lower order of complexity and ontologically separate; in Zeki's case, it is to identify the internalist proposal with the privileging of a specific level of reality (the neurophysiological one). The mirror image of this error is the endeavour to explain aesthetic meaning on the basis of phenomena of a higher order of complexity, such as social, linguistic, cultural ones or those of intersubjective distributed fruition (Duchamp 1958). Bottom-up components such as stimuli and neuronal correlates, and top-down components such as the influence of cultural or historic/social phenomena on the elaboration of pictorial and architectural styles, are elements that come into play at different levels in aesthetic complexity, but the latter is not reducible to them, neither in an externalist nor internalist sense.

The aforementioned aesthetic conceptions share several fundamental assumptions, such as:

1) Experiments are carried out using classic psychophysical or neuroscientific methodologies, which means beginning from the underlying features (be they stimuli or neuronal activity) instead of analysing and describing qualitative aesthetic-perceptual phenomena.
2) The explanations are identified and developed by devising models of a computational nature.

As varied as they may be, aesthetic theories of this kind generally overlook certain structural characteristics of the aesthetic perception of artworks; for instance, in the case of painting and sculpture:

1) The characteristic of *pure visibility*, i.e. the purely qualitative aspect of visual objects.
2) The specific nature of *spatial extension* as well as temporal extension of visual objects.
3) The nature and role not only of contents but also of *operations* implied in the act of vision.

In other writings (Albertazzi 2006a; 2006b) I have argued that certain laws of the perceptive organization of phenomenal space are similar to those of the organization of pictorial space, so that the analogy between many of the subjective processes involved at the primary level of phenomenal appearances and, for example, diverse modes of drawing on a canvas or carving a block of stone is more than a mere metaphor. That this is so is demonstrated by careful analysis of their structures and primitives, i.e. by the *appearances in visual space*. Despite their specificities, artworks are perceptual objects in which we identify lines, surfaces, volumes, colours and meanings just as we do in the vision of natural objects, and an accurate analysis of pictorial space often sheds light on phenomenal space and its characteristics of visibility.

By way of example I shall focus this contribution on the relationship between bi-dimensionality of surfaces and tri-dimensionality of volumes in 'natural' and 'pictorial' visual analysis, with especial reference to sculpture. In this way I hope to contribute to clarification of what is meant by *extended space outside of us*. The perspective adopted here is akin to Varela's but it maintains an ontological specificity of the experiential level that cannot be reduced to the underlying neurophysiological level. In contrast to the classic conceptions of externalism, it does not reduce perception to sensorimotor contingencies or a motor action theory, despite the fact that extended space clearly does have implicit cross-modal determinations of action (see below), and maintains a distinction between the various types of information that contribute to the formation of experience (Albertazzi, van Tonder and Vishwanath 2011). The concept of extended space affirmed here, the extendedness of which is not of a metric kind since it is essentially a space of operations of the perceiver, may form a bridge between the presently antithetical positions of externalism and internalism. This is the space of phenomenal appearances, bounded by forms themselves: indeed, appearances or visual objects are parts of space delimited by themselves from the inside.

The properties of the extended space can be shown by considering the simple but striking fact that in the visual field the perceptive evaluation of the *empty* distance between two points differs from the evaluation of a *straight* line drawn between two points, and also from the evaluation of the size, or better the *thickness*, of the two limit points, given that when a point has thickness it also acquires a *ground*. Of the same type are assimilative phenomena, like the fact that, if a square is placed between two squares both larger than it, and if it is then placed between two squares equally smaller than it, the squares appear to be smaller in the first case than in the second. A simpler example is the fact that the distance between two connected points appears shorter than that between two unconnected ones (already in Lipps 1897. On this topic, see Albertazzi 2002).

In what follows I shall therefore focus on bottom-up aspects of artistic perception connected with the phenomenal perception of artwork, necessarily omitting other aspects such as the epistemic dimension or cultural and historical-institutional aspects involved in the production of art and its appreciation (Danto 1983; 1997). Nor shall I deal with top-down cognitive or emotional processes related to artistic phenomena, such as those considered by philosophical theories on art of analytical derivation and closely connected with linguistic issues (Walton 1990; 2008). My approach is more radical also because it is based on a non-inferentialist theory of perception (Albertazzi, van Tonder and Vishwanath 2011).

3 Art and visual perception

My hypothesis, as upheld in other works (Albertazzi 2006a; 2006b; 2010, in press), may be summarized by the following points: as psychophysical organisms, we have *direct* access to the phenomenal perception of the environment, which is not reducible to the mere elaboration and transformation of stimuli by sensory organs and cerebral activity. There is much more in perception than just its stimuli, as shown by the structural 'ambiguity' of appearances, also because they are components of experience of a qualitative nature and with an intrinsic associated meaning (Pinna and Albertazzi 2011). All the information we receive from the environment around us is highly subjective, or rather subjectively integrated (Kanizsa 1991): it is characterized by secondary, tertiary, expressive qualities and dispositional qualities that are of great importance for both our survival and for the ecological perception of a scene. In fact, perception enables us to identify specific kinds of objects as well as to visually apprehend a wide variety of properties of the objects that go far beyond attributes understood in the computational sense like *metric* features or *cues*.

When looking *into* the surrounding environment (and *into* pictures as well; see Hecht, Schwartz and Atherton 2003), in fact, we do not 'see' a discrete series of points, or numbers representing curvature, distance, reflectance, etc. What we see are coloured appearances, *from which* we can define certain psychophysical measures such as slant, distance, orientation, and so on. In other words, the external measures (cues) are the ones that are 'parasitic' on the perceived entity that we for example call a surface, not *vice versa* (Vishwanath 2005). The figural primitives of these appearances are indeed lines, surfaces, volumes, but not metric cues; rather, they are operators of appearance, and of a qualitative nature—red, rough, luminous, transparent, etc.

The organization and different part/whole grouping of this kind of quality is the foundation of those perceptual configurations we call 'objects'—a tree, a statue, a stone, a painting, but also a shrill sound, a rough texture and so forth.

There are countless examples of how perceived reality only bears a feeble relationship with transcendent objects and stimuli—a conception put forward by Kant K.d.r.V, A387; Brentano 1874/1995; Helmholtz 1867; Hering 1920/1964—of which colour vision is a paradigmatic example (Bergström 2004), and of how the only realness for us is that of the space that extends ahead of us and all around us, of the objects we see near and far, that we are able to touch and manipulate, to hear and smell, and to integrate cross-modally. This is therefore not a metric space but an anisotropic space of forces, which is constructed by the perceiver and their operations (Lewin 1926; Arnheim 1954; Albertazzi 2006a).

From the Gestalt point of view, the distinction between inner and outer is a distinction 'internal' to the field itself. For example, the unilateral function of contour shows a segregation towards the inner area of the figure, and in ambiguous figures the same part can play the role of external or internal, albeit never simultaneously, and so on. At first sight it seems possible to speak of *different gradients* of externality and internality in the space of visual objects. In fact, the perceiver clearly plays a role as the origin of the system of spatial coordinates and thus determines the fundamental directions of space itself (so-called projection space). In this sense,

Koffka described the Ego as a region segregated and functionally different from others, or as a spatio-temporal structure embedded in a field (Koffka 1935).

From this point of view, the difference between natural phenomenal space and pictorial space is essentially a difference of continuity, rather than a categoric one. Indeed, both are capable of inducing a spatial representation, both are perceived spaces abundantly virtual with respect to the stimulus, sometimes so similar to each other so as not to be immediately distinguishable (such as in *trompe l'oeil*). Pictorial space is hence nothing but *a perceived space with a lesser degree of realness* (Metzger 1941).

If one excludes the perception not merely due to an elaboration of cues or metric/computational features and assumes instead that the construction of reality and the space of our senses is the result of a series of clues that the environment gives the perceivers and their sensory systems and that conscience organizes into a system of perceptual signs or configurations (in line with the conceptions of Berkeley 1709; Helmholtz 1867; see Koenderink 2011, in press), then the study of artworks becomes a true laboratory for the analysis of the laws of seeing and visual appearances. Indeed, the artist does nothing but test, exemplify and reshape the construction of appearances based on laws that are also active in the natural perception of objects. This point of view has also been expressed to some extent by, among others, Leonardo (see MacCurdy 1939), Klee (1961) and Kandinsky (1926).

To limit ourselves to visual art, as regards the characteristics of visibility and affordability, diverse artworks such as painting or sculpture share several laws that regulate the 'natural' phenomenal perception of surfaces and volumes. The intermodality that regulates natural perception (Spector and Maurer 2008) between colours and odours (Dematté *et al.* 2006), between colours and sounds, or between the lightness of colours and pitch (Marks 1978; Ward *et al.* 2006), should therefore have an analogue in artworks; one could then rule out a clear separation between sculpture (three-dimensional) and painting (two-dimensional) such as between vision and touch, despite their individual traits. Indeed, the sculptor begins with the motor representation that, as a whole, forms a visual image; the painter begins with the image from which to develop the plastic potential of visual impressions.

4 Surfaces and three-dimensionality

A fundamental aspect to our perceived quality of *realness* is that objects in a scene appear clearly volumetric and seem to 'stick out into space', appear touchable or graspable, etc., and the space between them is also perceived (Vishwanath 2011). This qualitative aspect of depth has been referred to as the plastic effect (Ames 1925; Judge 1926; Schlosberg 1941).

Today's prevalent point of view in vision science claims that the visual system is tasked with faithfully recovering the three-dimensional structure and layout of the visual world from the various sources of visual and non-visual information available, for example the retinal and extra-retinal depth cues. The visual system is meant to *recover* and *represent* two distinct properties of three-dimensional surfaces, their geometric structure and layout, and their associated pattern of reflectance (Adelson and Pentland 1996; Marr 1982; see Vishwanath 2011). This happens because, they maintain, the visual system's *internal representation* of a surface reflects the physical reality of surfaces in the *external world*. However, this explanation does not translate into the perception of pictures, where there simply is no three-dimensional surface layout (Vishwanath 2011).

Pictures are a special class of surfaces that, while being planar, have a complex pattern of coloration. This pattern of coloration automatically generates an illusory percept of depth and three-dimensional layout that contradicts the physical two-dimensional aspect of the stimulation (Vishwanath 2011). From a reductionist point of view, they are a main source of 'illusion', i.e. we see depth and three-dimensional structure where there is none. For these reasons (and related issues), pictures have become a privileged object of scrutiny in the scientific study of perception (Hagen 1980; Hecht, Schwartz and Atherton 2003; Kubovy 1986; Pirenne 1970; Wade, Ono and Lillakas 2001. See Vishwanath 2011). Indeed, although pictures are a type of visual surface rarely encountered in nature, the unusual effects that can be observed in pictorial perception offer clues towards the understanding of the perception of surfaces in general. A visual surface is an informational structure generated by the visual system that signals (affords, Gibson 1979) possible actions (Vishwanath 2011), and is therefore correlated with other modalities, such as touch and motive capabilities.

Both paintings and natural scenes therefore show how the realness of visual objects derives to a great extent from appearances made of *secondary clues* in extended visual space, of qualities such as flat, round, volumetric, soft, luminous, distant, close etc., and how the space of vision is essentially a space constructed by the observers and their *operations*, among which there are not only motor or tactile operations but the *act of seeing* itself as intrinsic to the visual system (see Cutting 2003; Hagen 1980; Koenderink and van Doorn 2003).

Obviously, we look *into* the space around us and *into* the space of pictures, sculptures or architectural monuments with a different sense of realness, of which we are contemporaneously aware. Indeed, we can perceive the brightness or the fogginess in a Turner sunset even though we are fully aware that we are not materially immersed in it, or we may visually perceive the softness in a Stevens painting even though we are fully aware of being unable to touch it (Koenderink and van Doorn 2003). The awareness of simultaneously participating in several nested phenomenal spaces offers several avenues towards an understanding of intermodality and diffuse synaesthesia of perception, as well as significant opportunities for the discussion of important topics such as why we see depth in pictures even when the predominance of visual *information* would suggest otherwise, and why pictorial depth so often appears to be *qualitatively different* from natural depth perception — unless one follows the aforementioned procedure.

As regards the two- and three-dimensionality of objects in vision, Michotte (1948/1991) was able to show experimentally that objects can appear in our perception with different degrees of phenomenal reality and cited this as a proof that our behavioural response differentiates 2D and 3D figures according to their perceptual salience. The experiment was conducted by asking the subjects to reproduce a real 3D object (a box) corresponding to the one represented. It showed that there was close correspondence among the subjects, and consequently that the perception of 3D was highly appropriate. A parallelepiped drawn on a sheet of paper, if observed from the top right, appears 'to stand up' (see Figure 1).

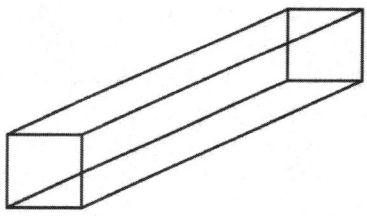

Figure 1: The parallelepiped appears to 'stand up', with one base attached to the paper and the rest of the body elevated in the space above the sheet.

The readers are invited to satisfy themselves that this happens when viewing with either one or both eyes. With both eyes, one must look at the paper at an 'extreme' angle, in which case the disparity information for the paper surface is very weak or disrupted (disparity gradient limit), so that one loses awareness of the paper surface, and the pictorial information dominates. This happens even more readily with one eye, where there is no conflicting binocular information. The impression of three-dimensionality is so strong that the observer willingly accepts an invitation to insert a thin rod in the 3D figure without considering the task meaningless, as it would be if s/he were looking at a normal 3D picture drawn on a similar sheet of paper.

The *qualitative* salience of the percept has been explained as being due to changes in assignment of egocentric distance information. The paper surface presence being disrupted, the distance information from accommodation might accrue to the perceived object allowing a scaling of the depth from an 'egocentric' point of view, thus giving it the quality of 'touchable' (Vishwanath 2011). The same does not happen, for example, if one draws a cube on a piece of paper that is viewed binocularly or only with slight slant, where distance information specifies the distance of the visual paper surface.

The *metric* two-dimensionality of figures therefore does not necessarily coincide with a *perceived* two-dimensionality and may even potentially provoke motor actions such as grasping the so-called illusory object. Analogous phenomena may appear in the vision of objects in pictorial space.

Because sculptures are visibly three-dimensional objects, they appear to manifest properties distinct from those of paintings, particularly ones of a tactile and motor nature. However, an analysis of the third dimension as a product of the observer's mind that idiosyncratically expands or contracts it (Koenderink and van Doorn 2003) shows that seeing and touching are not separate modes and, therefore, that the perfection of sculptures is also to be measured in their ability to render at a distance the three-dimensional volumetric qualities *with a two-dimensional surface effect* (Hildebrand 1893/1969). The conception and subsequent construction of a painting or a sculpture by an artist begins with an analysis of appearances and their visibility and is a development of this ability to sense the spatial dimensions based on the modes of visual and tactile sense that are themselves not separate modes. A modularity of senses can only be sustained by a strictly computational or reductionist neurophysiological conception of vision (Fodor 1983).

As an example of the theory I propose, I analyse the visual construction of the appearance of surfaces and volumes as presentations correlated with operations of seeing in extended space, with particular reference to pictorial space.

5 Seeing and touching

As previously mentioned, given the structural contiguity and continuity between perceptual and pictorial space, conceiving and assembling a painting or a sculpture may evidence certain operations that are also active in natural seeing. For instance, it is well known that the contents of two separate modes, such as seeing and touching, are different with respect to the same stimulus. Consider a visual phenomenon like a regular squared grid: in tactile perception (with 6- to 10-year-old children an experiment was performed and reported by Metzger 1975/2006), two rings with individual connecting lines emerge, in accordance with the law of common centre. If one then considers the child's drawing after touching, it shows the emerging of several shapes: for example, a sort of transitional form between the pattern of centre-surround organization, or frames with holes inside them, where the law of closure prevails.

Despite the difference in contents (Klatzky and Ledermann 2002), there exists a correlation between seeing and touching relative to the operations enacted in extended space, i.e. a correlation in *the way in which appearances are produced in sensory motions.*

An accurate aesthetic-epistemological analysis of this relationship between perceptual operations in vision and in the production of artwork is conducted in the aforementioned text by Hildebrand (1893/1969), under the influence of Helmholtz, which analyses the relationship between plastic form and appearance and its consequences for artistic representation. Hildebrand considers the operations carried out in the act of natural vision—i.e. the modes and conditions of visibility of appearances—and how these must be taken into account in the conception and realization of artwork because they represent the process of construction of appearance itself. Artwork thereby becomes an experimental laboratory of visual perception.

Hildebrand distinguishes between two kinds of visual activity, corresponding to two modes of optically perceiving objects, and identifies them within motor (essentially tactile) representation and unitary (essentially visual) representation. Motor representation has the near image as its visual correlate, while unitary representation has the far image. Appearance is constructed in the near image based on the various vistas of an object, and depth is inferred based on the comparison that may be established between visible surfaces: in this operation, seeing truly is a kind of touching, a tactile search carried out visually, and the complex of visual representations of lines and surfaces united by successive motive representations lies at the origin of plastic shape. A sort of tactile-visual phantasm occurs in this case, the content correlate of an actual cross-modal presentation.

On the other hand, in the far image every eye perceives only a surface, and depth can only be inferred from the contrasts on the apparent plane that act as clues for the reconstruction of an image of depth. The far image is thus the one that gives us unity of appearance.

The act of vision is therefore not the perception of an object's picture but rather the construction of a visual object in the form of a two-dimensional picture in the far vision. This process of natural vision is made plain and conscious in artistic representation. The difference between the production of a painting and a statue, for instance, lies in the fact that the 'material' with which the painter begins are the live impressions that s/he needs to express directly on a surface, giving the figure the appearance of a far image; a sculptor's 'materials' on the other hand are the motive representations derived from ocular movements and visual appearances. However, in order to be perfect a sculpture must be able to render a far image, despite its apparent volumetric features, i.e. an unitary image on the plane.

Some aspects of visual processes are fundamental for the realization of a far image: the movement that takes place in the act of seeing and the layout of spatial values based on a contextual relationship between parts and whole.

The (virtual) movement enacted in the operation of seeing and the construction of a far image is a movement that begins from the perceiver (recall Koffka's analysis, above) and continues in the spatial field in a backward direction, i.e. following a motion in depth. At this time, space *extends*, encountering the visual resistance of one- and two-dimensional appearances in the plane, a product of eye movement operations, until all spatial and shape relationships are unified in a single, simplified image: the far image.

The second aspect concerns the disposition of spatial values that are the result of subsequent contrasts, i.e. phenomena of grouping, figure/background, light-dark (i.e. shaded parts and lit parts), and of colour, which is an even more powerful carrier of distance than brightness values. Depending on the context, the various spatial values may have a greater or lesser weight, depending on the relationship between whole and parts. The relational ensemble of these contrasts contributes to the perception of a visual object as near or far.

The far image clearly gives the relationship between the dimensions of what lies on the plane in a relationship of contiguity. The unification of an image onto a plane, both in paintings and statues, takes place by means of occlusion, orientation and most of all in the unification of images in common planes of distance that avoid jumps and cuts in the virtual depth movement: in seeing a painting or a statue, the depth movement must be continuous as *the volume of a plane that continues in depth*. The image, or the shape, is the spatial *locus* wherein the act of seeing takes place (for the Aristotelian concept of locus see Albertazzi 2002).

The conception of the 'dual space' of the far and near image is supported by several scientific observations on the linkage of perceived realness (plastic shape) to changes in information on egocentric distance, and the related theory that perceived realness is the perception of the precision of the egocentric scale of space (Vishwanath 2011). Since egocentric scale (distance/size information) is most precisely available only in near space, the theory suggests that realism, and the corresponding sense of tactility, will be most manifest for near objects, declining continuously to the point where it is entirely absent in far space. Far space, where egocentric scale information is essentially abolished, thus becomes equivalent to pictorial space where egocentric scale information and the sense of *distance* is entirely ambiguous; unless it is introduced in the latter through special manipulations (e.g. the Michotte illusion, viewing through apertures – see Vishwanath 2011).

6 Aesthetics and science of qualities

As an analysis of space and perceived forms, aesthetics is a science as 'exact' as physics or biology: indeed, it is a theory of perception that may be experimentally verified. Some recent theoretical and experimental enquiries have led to the proposal of a physics of man (Koenderink and van Doorn 2003; Koenderink 2011, in press), the principles of which largely apply to aesthetics because they regulate perceptual operations and their contents, i.e. phenomenal appearances. The aesthetic analyses carried out from this point of view become a laboratory for a qualitative psychophysics: finding evidence of the laws of assembly of an image in various kinds of artwork by the artist, as well as shedding some light on the laws of seeing, may indeed suggest subsequent detailed experimental analyses to be carried out. The depth movement analysed by Hildebrand, i.e. the construction of a far image in which a figure's volume extends and is composed of subsequent homogeneous virtual planes until it reaches the stable state of an 'extended object', may for instance constitute a rich area for experimental enquiry. An analysis of this kind implies the identification of a precise theoretical hypothesis – that a specific 'geometry' of appearances exists and may be formally described – and of an adequate qualitative methodology that can 'do without the stimuli' for the analysis and the construction of a science of appearances. Research projects in this general direction, tending towards a science of qualities or the so-called 'measurement of the impossible', have recently been proposed, but they lack a comprehensive vision and an adequate theory and methodology. Experimental aesthetics contribute to bringing the qualitative aspects of our experience tied to perception, usually magnificently represented and expressed by the arts, away from the margins of scientific analysis where they have languished as if they were phenomena impossible to subject to measurement and to experimentation.

References

Adelson, E.H. and A.P. Pentland (1996), 'The Perception of Shading and Reflectance' in D. Knill and W. Richards, Eds., *Perception as Bayesian Inference*, New York, Cambridge University Press: 409–23.

Agrillo, C. and D. Roberson (2007), 'Colour Memory and Communication in English Speakers: Are Focals Easier to Name and Remember?' in A.L. Comunian and R. Roth, Eds., *International Perspectives in Psychology*, Aachen, Shaker Verlag Publishing: 79–88.

Albertazzi, L. (1997), 'Continua, Adjectives and Tertiary Qualities', *Axiomathes*, 8: 7–30.

Albertazzi, L. (2002), 'Towards a Neo-Aristotelian Theory of Continua' in L. Albertazzi, Ed., *Unfolding Perceptual Continua*, Amsterdam, Benjamins Publishing Company: 29–79.

Albertazzi, L. (2006a), 'Introduction to Visual Spaces' in L. Albertazzi, Ed., *Visual Thought: The Depictive Space of Perception*, Amsterdam, Benjamins Publishing Company: 3–34.

Albertazzi, L. (2006b), 'Visual Qualities. Drawing on Canvas' in L. Albertazzi, Ed., *Visual Thought: The Depictive Space of Perception*, Amsterdam, Benjamins Publishing Company: 165-94.

Albertazzi, L. (2010), 'The Ontology of Perception' in R. Poli and J. Seibt, Eds., *TAO-Theory and Applications of Ontology*, Vol. 1, Philosophical Perspectives, Berlin and New York, Springer: 177-206.

Albertazzi, L., G. van Tonder and D. Vishwanath (2011, in press), 'Information in Perception' in L. Albertazzi, G. van Tonder and D. Vishwanath, Eds., *Perception Beyond Inference: The Information Content of Perceptual Processes*, Cambridge (MA), MIT Press.

Ames, A. (1925), 'The Illusion of Depth in Pictures', *Journal of the Optical Society of America*, **10**: 137-48.

Arnheim, R. (1954), *Art and Visual Perception: The Psychology of the Creative Eye*, Berkeley (CA), The Regents of the University of California.

Bergström, S.S. (2004), 'The Ambeguias Phenomenon and Color Constancy', *Perception*, **33**: 831-5.

Berkeley, G. (1709), *New Theory of Vision*, Dublin.

Block, N., Ed. (1981), *Readings in the Philosophy of Psychology*, Cambridge (MA), MIT Press.

Boroditsky, L. (2001), 'Does Language Shapes Thought? Mandarin and English Speakers' Conception of Time', *Cognitive Psychology*, **43**: 1-22.

Brentano, F. (1874/1995), *Psychology from Empirical Standpoint*, L. McAlister, Ed., London, Routledge.

Chomsky, N. (1968), *Language and Mind*, New York, Harcourt, Brace and World.

Clark, A. (1998), 'Embodiment and the Philosophy of Mind' in A. O'Hear, Ed., *Current Issues in Philosophy of Mind, Royal Institute of Philosophy Supplement*, **43**, Cambridge, Cambridge University Press.

Clark, A. (2008), *Supersizing the Mind: Embodiment, Action, and Cognitive Extension*, Oxford, Oxford University Press.

Clark, A. and D. Chalmers (1998), 'The Extended Mind', *Analysis*, **58**: 10-23.

Crick, F. and C. Koch (1998), 'Consciousness and Neuroscience', *Cerebral Cortex*, **8**: 97-107.

Cutting, J.E. (2003), 'Reconceiving Perceptual Space' in H. Hecht, R. Schwartz and M. Atherton, Eds., *Looking Into Pictures: An Interdisciplinary Approach to Pictorial Space*, Cambridge (MA,) MIT Press: 215-38.

Danto, A.C. (1983), 'Art, Philosophy, and the Philosophy of Art', *Humanities*, **4** (1): 1-2.

Danto, A.C. (1997), *After the End of Art: Contemporary Art and the Pale of History*, Princeton, Princeton University Press.

Davidoff, J. (1991), *Cognition Through Colour*, Cambridge (MA), MIT Press.

Davidoff, J. (2006), 'Colour Terms and Colour Concepts', *Journal of Experimental Cognitive Psychology*, **94**: 334-8.

Davidson, D. (1980), *Essays on Actions and Events*, Oxford, Oxford University Press.

Dematté, M., D. Sanabria and C. Spence (2006), 'Cross-Modal Associations Between Odours and Colours', *Chemical Senses*, **31**: 531-8.

Duchamp, M. (1958), *Ecrits de Marcel Duchamp*, M. Sanouillet, Ed., Paris, Le Terrain Vague.

Fechner, T. (1876), *Vorschule der Ästhetik*, Leipzig, Breitkopf und Härtel.

Fodor, G. (1983), *Modularity of Mind: An Essay on Faculty Psychology*, Cambridge (MA), MIT Press.

Gallagher, S. (2009), 'Philosophical Antecedents of Situated Cognition' in P. Robbins and M. Aydede, Eds., *The Cambridge Companion to Situated Cognition*, Cambridge, Cambridge University Press.

Gibson, J.J. (1950), 'The Perception of Visual Surfaces', *American Journal of Psychology*, **63**: 367-84.

Gibson, J.J. (1979), *The Ecological Approach to Visual Perception*, Boston (MA), Houghton Mifflin.

Gregory, R. (1997), *Eye and Brain: The Psychology of Seeing*, Oxford, Oxford University Press.

Hagen, M.A. (1980), *The Perception of Pictures I: Alberti's Window: The Projective Model of Pictures*, New York, Academic Press.

Hecht, H., R. Schwartz and M. Atherton, Eds. (2003), *Looking Into Pictures*, Cambridge (MA), MIT Press.

Helmholtz, H. von (1867), *Handbuch der physiologischen Optik*, Leipzig, Leopold Voss.
Hering, E.E. (1920/1964), *Outlines of a Theory of the Light Sense*, L.M. Hurvich and D. Jameson, Trans., Cambridge (MA), Harvard University Press.
Hildebrand, A. (1893/1969), 'Das Problem der Form' in H. Bock, Ed., *Gesammelte Schriften zur Kunst*, Köln, Opladen, Bd: 39.
Honderich, T. (2006), 'Radical externalism', *Journal of Consciousness Studies*, **13** (7-8): 3-13.
Jackendoff, R. (1992), *Languages of the Mind: Essays on Mental Representation*, Cambridge (MA), MIT Press.
Johnson-Laird, P.N. (1983), *Mental Models*, Cambridge (MA), Harvard University Press.
Judge, A.W. (1926), *Stereoscopic Photography*, London, Chapman & Hall.
Kandinsky, W. (1926), *Punkt Linie zur Fläche*, Bern, Benteli.
Kanizsa, G. (1991), *Vedere e pensare*, Bologna, Il Mulino.
Klatzky, R.L. and S.J. Ledermann (2002), 'Tactile Object Perception and the Perceptual Stream' in L. Albertazzi, Ed., *Unfolding Perceptual Continua*, Amsterdam, Benjamins Publishing House: 147-62.
Kay, P. and W. Kempton (1984), 'What is the Sapir-Whorf Hypothesis?', *American Anthropologist*, **86** (1): 65-78.
Klee, P. (1961), *The Thinking Eye*, London, Lund Humphries.
Koffka, K. (1935), *Principles of Gestalt Psychology*, New York, Harcourt, Brace, and World.
Koenderink, J.J. (1998), 'Pictorial relief', *Philosophical Transactions of the Royal Society London A*, **356**: 1071-86.
Koenderink, J.J. (2011, in press), 'Vision and Information' in L. Albertazzi, G. van Tonder and D. Vishwanath, Eds., *Perception Beyond Inference: The Information Content of Perceptual Processes*, Cambridge (MA), MIT Press.
Koenderink, J.J. and A.J. van Doorn (2003), 'Pictorial space' in H. Hecht, R. Schwartz and M. Atherton, Eds., *Looking Into Pictures: An Interdisciplinary Approach to Pictorial Space*, Cambridge (MA), MIT Press: 239-99.
Köhler, W. (1929/1971), 'Ein altes Scheinproblem', *Die Naturwissenschaften*, **17**: 395-401. English trans. (1971), E. Goldmeyer, in M. Henle, Ed., *The Selected Papers of Wolfgang Köhler*, New York, Liveright: 125-41.
Kosslyn, S.M. (1980), *Image and Mind*, Cambridge (MA), Harvard University Press.
Kosslyn, S.M. (1994), *Image and Brain*, Cambridge (MA), MIT Press.
Kosslyn, S.M. and J.R. Pomerantz (1977), 'Imagery, Propositions, and the Form of Internal Representations', *Cognitive Psychology*, **9**: 52-76.
Kubovy, M. (1986), *The Psychology of Perspective and Renaissance Art*, New York, Cambridge University Press.
Lewin, K. (1926), *Vorsatz: Wille und Bedürfnis: mit Vorbemerkungen über die psychische Kräfte und Energien und die Struktur der Seele*, Berlin, Springer.
Lipps, Th. (1987), *Raumaesthetik und geometrisch-optische Täuschungen*, Leipzig, Barth.
Livingstone, M. (2002), *Vision and Art: The Biology of Seeing*, New York, Adam Press.
MacCurdy, E. (1939), *The Notebooks of Leonardo da Vinci*, New York, Reynal & Hitchcock.
Manzotti, R. (2009), 'No Time, No Wholes: A Temporal and Causal-Oriented Approach to the Ontology of Wholes', *Axiomathes*, **9**: 193-214.
Marks, L.E. (1978), *The Unity of the Senses. Interrelations Among the Modalities*, New York, Academic Press.
Marr, D. (1982), *Vision: A Computational Investigation Into the Human Representation and Processing of Visual Information*, New York, W.H. Freeman.
Massironi, M. (2000), *L'osteria dei dadi truccati*, Bologna, Il Mulino.
Maturana, H.R. and F.J. Varela (1980), *Autopoiesis and Cognition: The Realization of the Living*, Dordrecht, Reidel.
McManus, I.C., R. Cook and A. Hunt (2010), 'Beyond the Golden Section: Why do Individuals Differ So Much in Their Aesthetic Preferences for Rectangles?', *Psychology of Aesthetics, Creativity, and the Arts*, **4**: 111-26.
Merleau-Ponty, M. (1962/1945), *Phenomenology of Perception*, C. Smith, Trans., London, Routledge.

Merleau-Ponty, M. (1968/1964), *The Visible and the Invisible*, A. Lingis, Trans., Evanston (IL), Northwestern University Press.
Metzger, W. (1941), *Psychologie: die Entwicklung ihrer Grundannahmen seit der Einführung des Experiments*, Dresden, Steinkopff.
Metzger, W. (1975/2006), *Gesetze des Sehens*, Frankfurt, Kramer. L. Spillmann, Trans., 2006, Cambridge (MA), MIT Press.
Michotte, A. (1948), 'L'énigme Psychologique de la Perspective dans le Dessin Linéaire', *Bulletin de l'Academie Royale de Belgique*, **5** (34): 268-88. A. Costall, Trans., 1991, 'The Psychological Enigma of Perspective in Outline Pictures' in G. Thinès, A. Costall and G. Butterworth, Eds., *Michotte's Experimental Phenomenology of Perception*, Hillsdale (NJ), Lawrence Erlbaum Associates: 174-87.
Noë, A. (2004), *Action in Perception*, Cambridge (MA), MIT Press.
Noë, A. (2009), *Out of Our Heads: Why You Are Not Your Brain, and Other Lessons from the Biology of Consciousness*, New York, Hill & Wang.
O'Regan, J. and A. Noë (2001), 'A sensorimotor account of vision and visual consciousness', *Behavioural and Brain Sciences*, **24** (5): 939-1031.
Piattelli-Palmarini, M. (1980), *Language and Learning. The Debate between Jean Piaget and Noam Chomsky*, Cambridge (MA), Harvard University Press.
Pinna, B., Ed. (2008), *Art and Perception: Towards a Visual Science of Art* (Spatial Vision), Leiden, VSP.
Pinna, B. and L. Albertazzi (2011, in press), 'From Grouping to Meaning' in L. Albertazzi, G. van Tonder and D. Vishwanath, Eds., *Perception Beyond Inference. The Information Content of Perceptual Processes*, Cambridge (MA), MIT Press.
Pirenne, M.H. (1970), *Optics, Painting, and Photography*, Cambridge, Cambridge University Press.
Poli, R. (2001), 'The Basic Problem of the Theory of Levels of Reality', *Axiomathes*, **12** (3-4): 261-83.
Poli, R. (2006), 'Levels of Reality and the Psychological Stratum', *Revue Internationale de Philosophie*, **61** (2): 163-80.
Pylyshyn, Z.W. (1981a), 'Psychological Explanations and Knowledge-Dependent Processes', *Cognition*, **10**: 267-74.
Pylyshyn, Z. (1981b), 'Imagery and Artificial Intelligence' in N. Block, Ed., *Readings in the Philosophy of Psychology*, Vol. 2, Cambridge (MA), MIT Press: 170-94.
Ramachandran, V.S. (2003), *The Emerging Mind*, London, Profile Books.
Rock. I. (1983), *The Logic of Perception*, Cambridge (MA), MIT Press.
Rock. I. (1999), *Indirect Perception*, Cambridge (MA), MIT Press.
Schlosberg, H. (1941), 'Stereoscopic Depth From Single Pictures', *The American Journal of Psychology*, **54** (4): 601-5.
Searle, J. (1980), 'Minds, Brains, and Programs', *Behavioral and Brain Sciences*, **3**: 417-24.
Searle, J. (1983), *Intentionality: An Essay in the Philosophy of Mind*, Cambridge, Cambridge University Press.
Shannon, C.E. and W. Weaver (1949/1998), *The Mathematical Theory of Communication*, Urbana (IL), University of Illinois Press.
Sobchack, V. (2004), *Embodiment and Moving Image Culture*, Columbia and Princeton, University Presses of California.
Solso, R. (2003), *The Psychology of Art and the Evolution of the Conscious Brain*, Cambridge (MA), MIT Press.
Spector, F. and D. Maurer (2008), 'The Color of Os: Naturally Biased Associations Between Shape and Colour', *Perception*, **37**: 841-7.
Thompson, E. (2005), 'Sensorimotor Subjectivity and the Enactive Approach to Experience', *Phenomenology and the Cognitive Sciences*, **4** (4): 407-27.
Vishwanath, D. (2005), 'The Epistemological Status of Vision and Its Implications for Design', *Axiomathes*, **15**: 399-486.
Vishwanath, D. (2011, in press), 'Visual Information in Surface and Depth Perception' in L. Albertazzi, G. van Tonder and D. Vishwanath, Eds., *Perception Beyond Inference: The Information Content of Perceptual Processes*, Cambridge (MA), MIT Press.

Wade, N.J., H. Ono and L. Lillakas (2001), 'Leonardo da Vinci's Struggles with Representations of Reality', *Leonardo*, **34**: 231–5.
Walton, K. (1990), *Mimesis as Make-believe*, Cambridge (MA), Harvard University Press.
Walton, K. (2008), *Marvellous Images: On Values and the Arts*, Oxford & New York, Oxford University Press.
Ward, J., B. Huckstep and E. Tsakanikos (2006), 'Sound-Colour Synaesthesia: To What Extent Does It Use Cross-Modal Mechanisms Common To Us All?', *Cortex*, **42**: 264–80.
Zeki, S. (1999), *Inner Vision: An Exploration of Art and the Brain*, Oxford, Oxford University Press.

Part Two
Externalist Approaches to Aesthetics

Robert Pepperell

Art and Extensionism

There is an enduring, widespread, and deeply held belief that the conscious mind is located in the brain. One of the consequences of this belief is that many researchers who seek the causes of mental phenomena do so among the neural processes in the skull. As aesthetic experience is one kind of mental phenomenon it seems natural therefore to look to the brain if we want to explain it. The recent series of books, papers, and articles from eminent neuroscientists, psychologists, vision scientists, and others that apply knowledge from science to extend our understanding of art represent just such a 'neurocentric' approach. Foremost among this work is that of Semir Zeki, who can justifiably claim to be the prime mover behind the hybrid discipline of neuroaesthetics — a bold attempt to use neurobiological principles to account for subjective qualities in visual art, music, and taste.

In the Epilogue to what is perhaps the founding text of the new discipline, *Inner Vision*, Zeki (1999) states that what prompted him to write the book was a wish to '…learn whether there are any general statements that one can make about visual art in terms of what happens in the brain' (p. 217). His hope, vindicated as it turns out, is that '…looking at art as a product of the brain, through the workings of the brain and its functions, will continue' (*ibid*.) Subsequent texts such as those by Livingstone (2002), Ramachandran (2004), Solso (2003), Martindale *et al.* (2007), Skov (2009), as well as papers included in the *Journal of Consciousness Studies* series on 'Art and the Brain' (Goguen 1999), Spatial Vision's series on 'Vision Science and Art' (Pinna 2008), and talks delivered at several international conferences on neuroaesthetics all testify to the growing interest in accounting for art in broadly neurobiological terms.

But if we look across the disciplinary waters to the arts themselves we find, perhaps unsurprisingly, a rather different emphasis. Attempts to account for aesthetic experience by philosophers of art, art historians, art theorists, and artists themselves rarely make reference to what Zeki calls the 'strong biological foundations' of the process — a fact he notes. Perhaps this is out of understandable ignorance on the part of denizens of the arts about what goes on in brains, something neuroaestheticians may be able to rectify. But perhaps those who think about art from a 'native' perspective, as it were, simply do not presume artistic phenomena can be accounted for in any localized way. Unlike those trained in the physical sciences, who often instinctively seek local causal mechanisms for global behaviour, artists and art theorists—at least those I will be discussing here—look to the wider, systemic networks of cause and effect that bear on how artworks operate. These include things like the social and economic context within which a work is made and appreciated, the biographical factors that bear on its production, and the historical shifts it might precipitate or reflect.

Of course, with intellectual generosity these differing approaches— what we might call the neurologically bounded and the environmentally distributed—could be seen as complementary rather than conflicting. With much work it may be possible to graft them together into some fruitful union. I am of the view, however, that each approach reflects a fundamentally different understanding of the nature of the artistic experience being studied. This in turn rests on differing assumptions about where mental qualities are positioned in the world.

In the first part of this chapter I will discuss a range of ideas that in different ways suggest an understanding of art, and by extension the mind, that is unconfined to a specific location. In the second part I will present a thesis that explicitly denies a localized conception of experience. Instead I will propose a view of the mind and world as extended phenomena in which objects and events are distributed widely across time and space. The implications of this model for understanding art will be discussed.

1 Art as a distributed activity

The question of where artistic experience is located is actually a very old one, having been implicitly posed throughout the course of the long running debate in aesthetics between objectivists and subjectivists. Do qualities of beauty, taste, refinement, ugliness and so on reside in the object being contemplated or in the subjective experience of the contemplator? Earlier classical and pre-Enlightenment theories tended towards the former view, whereas Hume, Kant, and those following them tended towards the latter (Bunnin and Tsui-James 2003, pp. 232–3). Both objectivism and subjectivism, of course, presume a distinction between object and subject, and the inherent validity of this distinction is something we will address later. The concomitant belief that subjective experience occurs in the head while objective reality lies beyond will also be addressed.

In what follows I will cite various cases from art history and theory that resist a neat separation between the experience of art located inside in the head (or brain) and artwork situated separately in the world beyond. As we will see, they resist this separation not on ideological or purely philosophical grounds. Rather it is through trying to capture the very awkward essence of art—its peculiar combination of material artefact, conceptual stimulus, emotional impact, and cultural prestige—that certain thinkers have set their explanatory range far beyond local biological processes.

In *The Principles of Art* (1958) the philosopher R.G. Collingwood makes the claim that artistic activity is an aspect of conscious thought, consciousness for him being the agent by which sensuous experience is converted into imagination. Collingwood initially locates 'the work of art proper' as existing 'solely in the artist's head' (p. 305)—on the face of it a rather conventional subjectivist and neurocentric position. But in later defining aesthetic activity more fully he finds it necessary to acknowledge not only the role of the work of art in 'externalizing' the aesthetic sensibility of the artist but also the contribution of other artists whose influence is present in the work, as well as the collaborative role of the audience in the act of artistic creation. Thus he concludes by arguing against what he sees as the false view of 'individualistic psychology' in which artistic creation or appreciation is localized in one object or person: 'This [aesthetic] activity is a corporate activity belonging not to any one human being but a community' (*ibid.*, p. 324).

This communal conception of artistic activity, where the work is not confined to the artist, the art object, or the mind of the spectator alone, has been further explored by art historians. In *Only Connect... Art and the Spectator in the Italian Renaissance*, John Shearman (1992) describes how architectural spaces, sculptures, and pictures were organized by certain Renaissance artists so as to insist on the presence of the viewer in order to complete the work. In paintings such as the *Ecce Homo* panel by Quentin Massys (1515), now in the Prado in Madrid, the image is constructed so that the viewer becomes part of the depicted throng looking up at the condemned Christ on the balcony, and hence becomes complicit in his condemnation. For Shearman this is part of a historical shift towards what he terms 'the fully transitive work of art', which is a work 'whose subject is completed only beyond itself in the spectator's space, or even completed explicitly by and in the spectator himself' (*ibid.*, p. 59).

The art historian Ernst Gombrich's concept of the 'beholder's share', as presented in *Art and Illusion* (1960), refers to the way in which certain parts of a picture may be left empty or unresolved by the artist. The viewer then has the opportunity to supply the information necessary to complete the depiction from his or her own imaginative resources. Gombrich gives examples from Chinese brush painting, which combine economy of expression with deliberate omission, thereby encouraging the viewer to 'fill in' what is absent. Works of this kind raise the question of where precisely the artist action is occurring. The standard view, that it is entirely in the head of the spectator, risks underplaying the contribution made by the artist's mental resources, the way these are embodied in the work itself, and how these conspire with the viewer to produce the overall aesthetic effect. What Gombrich's carefully chosen phrase implies is that the beholder is responsible for providing precisely a *share*; that is, a portion of the total aesthetic workload. Other shares are vested, by implication, in the material of the work and the skill and imagination of the artist. Perhaps shares can be distributed even more widely, as Collingwood would have it, among the community of minds that constitute the culture in which the work is produced and appreciated.

The idea that aesthetic activity, or indeed any mental activity, can exist beyond the internal machinations of an individual person's head might smack of the archaic view of mind as soul. We are officially dismissive these days of inanimate objects harbouring spirits and thoughts being magically projected across space. Yet for much of human history beliefs of this kind were common, while in numerous cultures across the world they are still publicly endorsed. In *Totem and Taboo*, a collection of essays concerning 'Points of Agreement between the Mental Lives of Savages and Neurotics', Sigmund Freud (1991) uses the phrase 'omnipotence of thought' to refer to various uncanny, telepathic, and superstitious ways of thinking, where thought is seen to extend beyond the individual to affect people and objects without normal physical transmission. Freud identifies these beliefs with primitive stages of development, having been superceded in what he regards as civilized culture and the mature individual—with one exception:

> In only a single field of our civilization has the omnipotence of thoughts been retained, and that is in the field of art. Only in art does it still happen that a man who is consumed by desires performs something resembling the accomplishment of those desires and that what he does in play produces emotional effects—thanks to artistic illusion—just as though it were something real. (Freud 1991, p. 90)

Freud goes on to say that the primal artistic impulse was founded in the magical principle that to represent something was to conjure it up, to control it, so that it might serve as a material extension of one's thoughts and desires.

Anthropologists have made extensive studies of cultures in which the belief in the capacity of inanimate objects to carry mental impulses is central to social organization. The first to fully develop an anthropological theory of visual art was Alfred Gell, whose book *Art and Agency* (1998) looks at the way artefacts function as external manifestations of internal conscious states. Designs and objects made by the Marquesan peoples of Polynesia are examined to show how a particular society's artworks can form a distributed corpus, where each object relates to the other and all relate to the collective whole: '...each fragment of Marquesan art resonates with every other, because each has passed, uniquely, through a Marquesan mind, and each was directed towards a Marquesan mind' (*ibid.*, p. 221). He goes on to argue from this: '...there is an *isomorphism of structure* between the cognitive processes we know (from inside) as "consciousness" and the spatio-temporal structures of distributed objects in the artefactual realm... the structures of art history demonstrate an externalized and collectivized cognitive process' (*ibid.*, p. 222). Gell develops this into a more general thesis of the 'extended mind' that transcends 'the individual *cogito* and the coordinates of any particular here and now' (*ibid.*, p. 258).

One of the modern European artists Gell cites in support of his thesis is Marcel Duchamp, whose *oeuvre* he regards holistically as an interconnected treatise on the 'continuum' of reality beyond the material, what contemporaries of Duchamp commonly referred to as the 'fourth dimension' (*ibid.*, p. 243). Gell argues that works like *The Large Glass* of 1913–25 and *The Network of Standard Stoppages* of 1914 are part of a 'coherent project, which... subsequently extended itself until the close of his career' (*ibid.*, p. 245). These works form a 'single distributed object' the purpose of which was not so much to represent the elusive reality beyond appearances but to 'create a fourth-dimensional entity' through the work itself. The result, Gell argues, is that Duchamp's very subjectivity is instantiated in a series of creative moments, each represented by a different work:

> In other words, as a distributed object, Duchamp's consciousness, the very flux of his being as an agent, is not just 'accessible to us' but has assumed this form. Duchamp has simply turned into this object, and now rattles around the world, in innumerable forms, as these detached person-parts, or idols, or skins, or cherished valuables. (*ibid.*, p. 250)

For his part, in a brief but famous lecture entitled 'The Creative Act', given to the Convention of the American Federation of Arts in Houston in 1957, Duchamp set out a theory of artistic creativity in which the 'role of the spectator is to determine the weight of the work on the aesthetic scale', whatever the prior hopes or intentions of the artist might have been (in Lebel 1959, pp. 77–8). Like others cited here, Duchamp resists restricting aesthetic experience to any specific location, seeing it instead as a consequence of distributed activity among several players:

> All in all, the creative act is not performed by the artist alone; the spectator brings the work in contact with the external world by deciphering and interpreting its inner qualification and thus adds his contribution to the creative act. (*ibid.*)

These oft-quoted lines of the artist are sometimes cited as theoretical justification for the conceptual approach to art-making, substantially initiated by Duchamp through his cycle of 'ready made' art works, such as *Bottle Rack* (1914), *In advance of the broken arm* (1915), and *Fountain* (1917). As its name implies, conceptual art emphasizes the interpretative aspects of artistic engagement over the formal visual or visceral aspects (although one might question the ultimate validity of such a distinction). With artworks made from objects chosen, as Duchamp claimed of his bottle rack, because of their complete lack of aesthetic merit, the onus is then on the viewer to supply the judgment and interpretation necessary to complete the creative act, and thus create the work of art. Under this Duchampian paradigm, like other cases noted above, the aesthetic activity is not located in a specific place but distributed between the artist, artwork, and the spectator.

Implicit in all such exchanges, and central to the interventions Duchamp made through his ready-mades cycle, is the wider institutional framework of the art world. It was not so much the selection of coat hooks and urinals as works of art that unsettled as much their deliberate introduction into the gallery context, outside which of course they were entirely unremarkable.

The problem of how to classify objects as works of art and separate them from objects that aren't consumed art theorists and critics for much of the last century, and continues to do so today. It was a problem exacerbated by the determination with which artists made works intended to confound any attempt at definition in purely formal terms. Duchamp had already undermined the assumption that works of art have unique formal properties by maintaining that anything placed in a gallery can claim artistic status. Eventually the gallery context itself was displaced by a whole series of art movements that emerged in the 1950s, 60s and 70s, including various kinds of situated, performance, installation, landscape, and interventionist practices that avoided being subsumed into traditional art venues.

One response to this fragmentation was to redefine art in terms of its wider institutional dependence rather than its formal physical properties or particular spatial location. A prominent advocate of this so-called 'institutional theory of art' was George Dickie (1974) who proposed that a work of art is an object '...the set of aspects of which has had conferred upon it the status of candidate for appreciation by some person or persons acting on behalf of certain institutions (the artworld)' (*ibid.*, p. 34). The locus of aesthetic activity then shifts from the object itself, even beyond the artist, spectator, or gallery to the much wider network of arts organizations, museums, art schools, dealers, publications, critics, etc. who collectively constitute the socially sanctioned art world. Under this definition we can no longer appreciate the artwork on an individual basis, or within its immediate local context, but instead as an object that stands in for the accumulated actions of many disparate agents whose judgment it either implicitly or explicitly represents.

In some ways the offspring of this institutional approach to defining art practice are the various forms of so-called relational art that emerged on the 1990s European art scene. These are process-led, socially engaged works that include, for example, Rikrit Tiravanija's dinner party in a collector's home, or Christine Hill's piece in which she works as a check-out assistant and organizes a weekly gym workshop in a gallery. These and other examples are discussed by the art critic Nicolas Bourriaud in his influential collection of essays *Relational Aesthetics* (2002). Relational art, as Bourriaud defines it, takes '...as its theoretical horizon the realm of human interactions and its social context, rather than the assertion of an independent and *private* symbolic space' (*ibid.*, p. 14). The intersubjective 'encounter' between artwork and beholder, he argues, has always been part of the function of art: 'art has always been relational in varying degrees' (*ibid.*, p. 15). But in the new work Bourriaud sees the social engagement not as the consequence of contemplating an art object but as the artwork itself.

> The artistic practice thus resides in the invention of relations between consciousness. Each particular artwork is a proposal to live in a shared world, the work of every artist is a bundle of relations with the world, giving rise to other relations, and so on and so forth, ad infinitum... As part of the 'relationist' theory of art, intersubjectivity does not only represent the social setting for the reception of art... but also becomes the quintessence of the artistic practice. (*ibid.*, p. 22)

The culture of interactivity that coincided with the emergence of relational art was largely precipitated by the video games and communication technology explosion of the 1990s. Mobile phones, email, digital cameras, social networking sites, and online communities started to palpably affect the way members of (mainly developed) societies engaged with each other, and more particularly the way images were made and consumed. This led some art theorists to argue that a fundamental shift had occurred in our relationship to technology and the images that now flowed through it at unprecedented quantity and speed. Artist and new media theorist Ron Burnett, for example, made the case in *How Images Think* (2004) that emerging modes of collaboration between humans and machines, such as remote co-working, peer-to-peer communication, and networked musical composition, meant human intelligence had become a distributed phenomenon, whereas once it was confined to individual sentient beings. The media with which this collaborative activity is imbued—the sounds, texts and, most importantly for Burnett, the images—then gain a kind of intelligent status of their own. Technologically-mediated images acquire cognitive attributes, and so '...turn into intelligent arbiters of the relationships humans have with their mechanical creations and with each other' (*ibid.*, p. 221).

For artist and theorist Roy Ascott, meanwhile, the new modes of social engagement afforded by mass-participation online worlds like Second Life, and the variable personas we can adopt when inhabiting them, present new artistic possibilities. He sees the dissolution of the unique self in what he calls the 'variable reality' of cyberspace, where both individual and environment are much less stable and monolithic than in conventional reality.

> The real technological revolution in art and in society lies not simply in this global connectivity of person-to-person, mind-to-mind (significant as that is), but in its power to provide for the release of the self, the fictive 'unified self' of Western philosophy. (Ascott 2009, p. 24)

What such virtual realities offer for '...artists in search of new insights, images, systems and structures, new intellectual, social and spiritual associations and relations...' (*ibid.*) is a highly distributed, responsive, and variegated environment in which the traditional notion of a unique individual mind located in a specific place recedes. The kind of art that will populate this variable reality will induce new forms of aesthetic engagement and indeed, according to Ascott, new kinds of consciousness. The displacement of presence, the extension of mind, the accumulation of knowledge, the amplification of action, and the global synchrony afforded by near-instant communications all suggest some deep and lasting change is underway in the constitution of our mental life. With it will inevitably come a commensurate alteration in our age-old impulse to make, share, and consume works of art.

The examples presented here represent a way of thinking about art and aesthetic experience that cannot be reduced or isolated to a specific location, least of all the brain or some part thereof. This is not to deny, of course, that a functioning brain is essential to our ability to fully engage with the world, and with works of art as we encounter them. But the notion, implicit in neuroaesthetics, that we can understand something about the nature of art by looking how the brain responds under certain conditions is too limiting; it may tell us something about the brain but little about art. This is because what we have been referring to here as 'aesthetic activity' (to borrow the phrase from Collingwood) always consists in more than what occurs in part of any individual person. Properly understood artistic activity is a highly distributed process that exists as a function of numerous minds, diverse materials and technologies, countless creative actions, all occurring within a socially structured environment. All these factors conspire simultaneously to produce the overall artistic experience.

2 Extended minds and objects

If we pose the question 'Where is the artwork?' in relation to, say, a painting by Picasso, the common sense reply would be that it is coincident with the material form in which it is encapsulated. And if we were to ask 'Where are the edges of the artwork?' we would expect a response along the lines that it ends at the physical extremities of its matter, that is, at the local surface of the paint or the side of the frame, or wherever. All this seems on the face of it quite straightforward, but I want to suggest there is a rather more awkward ontological problem underlying these common sense responses. If the artwork ends at the local extremity of its physical structure then it is fair to say it ceases to exist beyond that point. If the outer skin of the paint constitutes the extent of the work then the space immediately surrounding it does not. By the same logic, we as observers must cease to exist at the outer extremity of our own skin if we are only constituted by what lies inside it. Assuming we are looking at the Picasso from a distance of a metre or so, which would be normal in most gallery settings, there is a gap (I'm tempted to call it an explanatory gap) between ourselves and the work on the wall in which nothing of it, or us, exists. How is it then we can have any contact with an object that doesn't intersect with us in any way?

We can resolve this immediately by appealing to the simple action of light. We see the Picasso only because the light in the room (from the lamps or skylights, or whatever) is reflected from the painting's surface into our eyes. There then follows a complex process, more or less well understood, involving certain optical laws and neurobiological activity, at the end of which we have the experience of seeing the work. All this would be fine if it were not for the fact that we have, in a sense, 'lost sight' of the Picasso. We can no longer claim we are experiencing the object itself but the patterns of light it reflects. But surely art thieves would not be content to steal these patterns of light; patterns of light can't be auctioned; they have no artistic or commercial value in themselves. Yet as we stand before the work it is only through these patterns that we have a direct experience of the work at all.

To settle this problem let us reject the idea that the artwork ceases to exist at the point where the outer skin of the paint ends. Let us grant that the patterns of light reflected by the Picasso are in some sense constituent parts of the work. After all, they are unique to this particular painting. It is these patterns of light that Picasso effectively manipulated via his treatment of the paint as he applied it, and which he intended us to experience as viewers. These same patterns (or an approximation of them) are evident when the painting is reproduced, copied or forged; they form a kind of visual fingerprint by which the image can be exclusively identified; nothing else but this work would produce them in quite the same way. It seems reasonable then to say they are part of the painting — part of what makes it what it *is*.

Allowing that the patterns of light are a constituent part of the work helps us avoid the 'gap' conundrum; our sensory systems are now in direct contact with something that is entirely dependent on the work; the Picasso has been restored to our field of view. But if we were to repose the question about where the work is located we get a rather different answer. With the patterns of light now being part of the work the spatial boundary of the object has expanded from the immediate surface to the furthest extent of wherever the reflected light travels. Some of it hits our eyes, for sure, but much of it travels further afield, bouncing into all kinds of other surfaces, being deflected in numerous and probably incalculable ways until absorbed or converted into energy of some other kind. As a consequence the physical extent of the work has become rather large, perhaps indefinitely large.

This simple case study is intended to make an important point bearing on the question being addressed in this chapter of how to characterize the relationship between art and the mind. The assumption that objects, including art objects, are delimited by their apparent surface boundaries is, when looked at from another perspective, more nonsensical than commonsensical. For if objects were confined only to the extent of their local boundaries there would be no means of contacting them beyond those boundaries, and we could have no experience of them of the kind we habitually have. By allowing that the patterns of light reflected from an object are constituent parts of the object we have restored the common sense link between observer and observed. Objects can now be thought of as being made up in part by the features they present to us, such as the light they reflect, or in the case of a sound piece the audio waves it sends rippling through the air, or the tactile sensations induced in the nervous system by a touch-based piece, or (to move to the culinary domain) the taste a particular food engenders in our gustatory system through the chemical compounds it releases when chewed.

The general principle is this: that objects are constituted as much by their dependent properties (these being any physical properties that depend on the object for their existence) as they are by what lies within their apparent local boundaries. Anyone wanting to insist on the obvious objection, that the pattern of light from a Picasso painting is not to be confused with the painting itself, must then give an alternative account of how it is through this pattern and only this particular pattern that the painting is known to us as we look at it in the gallery. And anyone who concedes the patterns of light are constitutive of the painting, but objects that this is only marginally so on the grounds that the 'real' substance of the painting is the material object hanging one metre from our eyes would be denying the reality of our experience as we look at the painting, which is based almost entirely on just those patterns of light.

If we extend this principle and consider that the painting as an object is dependent for its existence on the artist, in this case Picasso, then we must allow that the painting is really a part of Picasso. Some may find this harder to swallow since it suggests that a now-dead person somehow continues to exist in the form of an inert object made of humble canvas, paint and wood. But if we give a little thought to the processes involved it is actually rather uncontroversial. As noted, in the same way that we cannot restrict the constitution of a painting to its local physical boundaries we cannot restrict the constitution of a person to their skin. If nothing of a person ever reached beyond this barrier we would know little about each other at all. I would suggest a particular person is comprised not only by their local physical frame but also their actions, behaviours, habits, expressions, and all the various effects they have on the environment around them. The unique marks we make on the world are part of our existence taken as a whole. They, like the air, liquid and other matter we excrete, our voices, gestures, smells, signatures, are what we might call 'extended' aspects of our being—and no less part of us for being extended.

The case of artists in this regard is especially interesting. Part of the value, prestige and excitement surrounding a work by an artist of the stature of Picasso is the uniqueness of its attachment to him and the degree of proximity it affords to his person. Prints or paintings that lack Picasso's personal signature are generally far less valuable than those that have it (which is why so many unscrupulous dealers have become experts at signing his name). Why? Not because the signature makes the image much better on aesthetic grounds, but because it is evidence of the personal attention the artist has paid to the particular object he signed, and of its authenticity. This is why it is sometimes said it is the artist who is bought rather than the artwork; many works of art are valuable not so much of what they are (the raw material value of most great works of art is negligible) but because of the person they are associated with. (As I write, a rather undistinguished blue period Picasso has sold for a record sum.) They are part of, in a real sense and not simply in a metaphorical sense, what the artist was as a person and how they lived their life. Their being extends into the work, and through the work we come into contact with this aspect of their being. To own a Picasso is to make Picasso part of one's own.

The use of the word 'being' in this context is intended to capture both the physical aspects of a person's activity in the world as well as their mental disposition, that is, their experience, consciousness, cognition, memories, emotions, and so on. Artworks are indeed for the most part physical objects. But a large part of the reason they are made and appreciated is their capacity to capture, express, and stimulate certain mental states. We might even go as far as Gell in seeing art objects as structurally isomorphic with the consciousnesses of the artists who make them, and by extension the audiences that engage with them. In this sense, then, the artwork becomes a literal extension of the mind of the artist, embodying their unique cognitive attributes as well as the physical actions they employed in its construction. Our own conscious engagement with the work therefore links the mind of the artist to our own. Part of the thrill of seeing a great work is the vivid sense one gets of an organizing intelligence manifest in the physical form of the work, an intelligence that is reconstituted as part of one's own mind in the process of contemplation. Appreciating a work of art is somehow to resonate with the mind of the person that made it, even is that person is long dead.

I am presenting here a way of thinking about mind, reality, and being that elsewhere I have termed 'extensionism' (Pepperell 1995/2003; 2005). Simply put, extensionism stresses the continuities between objects and events rather than the distinctions. Where a conventional analysis might assume that a certain object or event was discretely bounded in some way—by having a visible edge, or displaying a change over time—an extensionist analysis would look instead for the continuity between the edge and what it abuts, or the way in which we are required to maintain awareness of a prior state in order to be aware of a new event. Extensionism does not deny or preclude that objects and events can *appear* to have edges, boundaries or limits, or that in the course of everyday life we all take such boundaries for granted as we move around the world. Boundaries and distinctions are 'real' as much as anything is real in our experience of the world. But they are not mind-independent properties that exist whether or not anyone is there to perceive them (an application of the concept 'real' I know some will resist). Boundaries and distinctions emerge as a consequence of the way our perceptual and cognitive systems divide and categorize sensory input consistent with the biological needs of the organism.

Extensionism takes as its starting point the idea that all objects and events have extended dimensions, most of which we remain unaware of most of the time. By extended dimensions I mean properties that are uniquely associated with the object or event in question—what I referred to above as dependent properties. These might include the history of the material from which an object is made, its place in a chain of social signification, its links with all the people who have ever come into contact with it, its kinship with other similar objects, its formal evolution, the intellectual or creative energy it embodies, its place in the gravitational field, and so on. We are aware of some of these properties, more or less dimly, each time we encounter an object or experience an event. But every object and event has far more dependent properties than we could ever contemplate; in each case they ripple indefinitely through space and time, connecting, however slightly, to countless other objects and events unknown. The sheer practicalities of life prevent us from acknowledging all but a fraction of the extended dimensions of each apparently discrete thing we know.

Extensionism is not a unique or even recent idea. In fact, it is a rather venerable way of looking at the world that has resurfaced in various forms over the epochs. An early manifestation can be found in the Buddhist doctrine of 'dependent origination' dating from around 500 BC. Variously expressed, it holds that no object can exist in isolation from everything else, and each object is in some way dependent on or conditioned by other objects. The fact that consciousness, acting on sensory data, breaks up the world into individual entities, some of which we feel isolated from and some of which we crave, leads us into a false relationship with reality. By recognizing the truly integrated nature of all seemingly distinct things, including ourselves, Buddhists argue, we can avoid the sense of dislocation, alienation, and suffering that we would otherwise have to endure (Conze 1962, p. 187). As with the extensionist view of reality, under the doctrine of dependent origination objects have no discrete boundaries other than those we impose upon them from our peculiarly human perspective.

There is also an affinity between extensionism and some of the views propounded by Alfred North Whitehead, especially his notion of the 'fallacy of simple location' of objects. In *Science and The Modern World* (1967), Whitehead argues that science, especially physics, has operated for too long on the basis that reality is made up of isolated, discrete chunks of matter that have a fixed locations in space and time. The result has been to perpetuate the illusion whereby we perceive objects as though they were independent entities:

> ...my theory involves the entire abandonment of the notion that simple location is the primary way in which things are involved in space–time. In a certain sense, everything is everywhere at all times. For every location involves an aspect of itself in every other location. Thus every spatio-temporal standpoint mirrors the world. (*ibid.*, p. 91)

Whitehead says that although this may appear paradoxical it is in fact entirely obvious when we consider how it is that we actually perceive objects in the world, which are never in isolation, and are always connected, always '...fading away into the general knowledge that there are things beyond' (*ibid.*, p. 92).

Philosophers often distinguish between an object's primary properties, i.e. its 'objective' features such as weight, size, volume, etc. and 'secondary' qualities, which are those that depend on the way it appears to a person depending on their disposition. In the case of the Picasso, the patterns of light reflected from a surface would be seen as secondary properties of the painting as they will appear different to each spectator. The temptation is to regard the primary properties as constituting the 'real' painting and the secondary properties as somehow ephemeral and less consequential; they fall easy prey to what Whitehead calls 'subjectivist criticism'. The essence of extensionism, however, is to transcend this distinction and treat secondary properties as a proper part of the wider constitution of an object. Whitehead agrees, but notes how difficult this can be because we are so used to thinking about an objective world made of discrete primary-level realities behind a subjective world of secondary qualities. 'If we are to include the secondary qualities in the common world, a very drastic reorganization of fundamental concept is necessary' (*ibid.*, p. 91). It is precisely this kind of reorganization of our habitual beliefs that the extensionist approach seeks to promote.

Finally, extensionism is consistent with the various extended mind and distributed cognition theses to have emerged in science and philosophy over recent decades, such as those of Gregory Bateson (1972), Edwin Hutchins (1995), Clark and Chalmers (1998), Max Velmans (2003), Francois Tonneau (2004), Rupert Sheldrake (2005), Ted Honderich (2006), Andy Clark (2008), Alva Noë (2009), Riccardo Manzotti (2009), and others. Broadly speaking, these theses take the view that certain mental properties, such as memory, belief, creativity and intelligent calculation, can depend on activity beyond the immediate locus of the individual brain, which as we saw at the outset of this chapter is where such properties are traditionally thought to reside. The philosopher Alva Noë, for example, in *Out of Our Heads* (2009) contends that:

> You are not your brain. We are not locked up in a prison of our own ideas and sensations. The phenomenon of consciousness, like that of life itself, is a world-involving dynamic process. We are already at home in the environment. We are out of our heads. (p. xiii)

Noë's view has implications for the way we relate to art:

> Just as we do not draw an impermeable boundary around the brain, we will not draw such a boundary around the individual organism itself. The environment of the organism will include not only the physical environment but also the habitat, including, sometimes, the cultural habitat of the organism. (*ibid.*, p. 185)

Extensionism, then, as presented here, applies these same integrative principles to objects and events as well as minds, and can be summarized in the following claim:

All objects and events have extended (non-local) dimensions, but we normally acknowledge only a fractional part of their true extent because of constraints inherent in our perceptual apparatus and the coercive effects of time. Rather than regarding discernible objects and events in the world as integral and discrete we must recognize that they, their origins and their repercussions extend indefinitely through space and time.

The experience of looking at a work of art, as described here, does not assume an essential division between the external object and the internal subjective mind of the viewer. Rather, one extends to the other, forming a continuum in which the mind reaches out to the work as much as the work reaches into the mind. In this way the mind, the work, and indeed the artist, become fused. The depth and richness of this fusion — what we might call the level of aesthetic experience — is determined by the skill and intelligence invested by the artist in the artwork and the receptivity of the viewer in interpreting that skill and intelligence. This fusion and all the mental properties that go to make it up have no simple location, to use Whitehead's phrase. They are distributed in time and space, woven at every level into myriad other objects and events; they are functions of minds that extend far beyond space of any brain or any immediate present.

3 Conclusion

This chapter took as its starting point a critique of the assumption implicit in the hybrid discipline of neuroaesthetics that we can look to the brain to explain artistic experience. This, of course, is an assumption predicated on the belief that the brain is the 'seat' of consciousness. I have argued that although looking at what happens in the brain may give us insights into certain restricted parts of the process, to fully understand the way works of art have their effect on us is to appreciate the extended, non-local ways in which they function.

On the back of this a further claim has been made: that what applies to aesthetic activity applies also to the operation of the mind more generally. Just as we cannot limit our consciousness of a work of art to a specific region of neurospace, so we cannot locate consciousness in any precise region of the body, or world. The brief exposition of the extensionist approach offered here takes an ontological stance in which our view of reality as being made up of numerous discrete and delimited entities with an existence that is independent of consciousness is seen as at best pragmatic, at worst fallacious. Nor can we any longer accept a fundamental firewall between our subjective experience of objects and objects in themselves. A richer and deeper understanding of our nature and the nature of the world (which amount to the same thing) requires we acknowledge the limitless extent of all things we take as localized, including the mind itself. This means in a certain sense to be conscious of something *is to be that something*; the mind extends to all things we are mindful of. Much as this runs counter to our habitual modes of thought, like the genie freed from the bottle, once out it cannot be returned to its earlier confinement.

References

Ascott, R. (2009), 'The Ambiguity of Self: Living in a Variable Reality' in R. Ascott *et al.*, Eds., *New Realities: Being Syncretic*, Vienna, Springer-Verlag.
Bateson, G. (1972), *Steps to an Ecology of Mind: Collected Essays in Anthropology, Psychiatry, Evolution, and Epistemology*, Chicago, University Of Chicago Press.
Bourriaud, N. (2002), *Relational Aesthetics*, Paris, Les Presses du Réel.
Bunnin, N. and E.P. Tsui-James (2003), *The Blackwell Companion to Philosophy*, Oxford, Blackwell.
Burnett, R. (2004), *How Images Think*, Cambridge (MA), MIT Press.
Clark, A. (2008), *Supersizing the Mind: Embodiment, Action, and Cognitive Extension*, Oxford, Oxford University Press.
Clark, A. and D. Chalmers (1998), 'The Extended Mind', *Analysis*, **58**: 10–23.
Collingwood, R.G. (1958), *The Principles of Art*, Oxford, Oxford University Press.
Conze, E. (1962), *Buddhist Thought in India*, London, George Allen & Unwin.

Dickie, G. (1974), *Art and the Aesthetic: An Institutional Analysis*, Ithaca (NY), Cornell University Press.
Freud, S. (1991), *Totem and Taboo*, London, Routledge.
Gell, A. (1998), *Art and Agency: An Anthropological Theory*, Oxford, Oxford University Press.
Goguen, J., Ed. (1999), *Art and the Brain*, Exeter, Imprint Academic.
Gombrich, E. (1960), *Art & Illusion: A Study in the Psychology of Pictorial Representation*, London, Phaidon Press.
Honderich, T. (2006), 'Radical Externalism', *Journal of Consciousness Studies*, **13** (7–8): 3–13.
Hutchins, E. (1995), *Cognition in the Wild*, Cambridge (MA), MIT Press.
Lebel, R. (1959), *Marcel Duchamp*, New York, Pangraphic Books.
Livingstone, M. (2002), *Vision and Art: The Biology of Seeing*, New York, Abrams Press.
Manzotti, R. (2009), 'No Time, No Wholes: A Temporal and Causal-Oriented Approach to the Ontology of Wholes', *Axiomathes*, **9**: 193–214.
Martindale, C., et al. (2007), *Evolutionary And Neurocognitive Approaches to Aesthetics, Creativity And the Arts*, London, Baywood.
Noë, A. (2009), *Out of Our Heads: Why You Are Not Your Brain, and Other Lessons from the Biology of Consciousness*, New York, Hill and Wang.
Pepperell, R. (1995/2003), *The Posthuman Condition*, Oxford, Intellect (revised as *The Posthuman Condition: Consciousness Beyond the Brain*, 2003, Bristol, Intellect).
Pepperell, R. (2005), 'Posthumans and Extended Experience?', *Journal of Evolution and Technology*, **14**.
Pinna, B., Ed. (2008), *Art and Perception: Towards a Visual Science of Art* (Spatial Vision), Leiden, VSP.
Ramachandran, V.S. (2003), *The Emerging Mind*, London, Profile Books.
Shearman, J. (1992), *Only Connect... Art and the Spectator in the Italian Renaissance*, Princeton, Princeton University Press.
Sheldrake, R. (2005), 'The Sense of Being Stared At: Is it Real or Illusory?', *Journal of Consciousness Studies*, **12** (6): 10–31.
Solso, R. (2003), *The Psychology of Art and the Evolution of the Conscious Brain*, Cambridge (MA), MIT Press.
Skov, M. (2009), *Neuroaesthetics*, London, Baywood.
Tonneau, F. (2004), 'Consciousness Outside the Head', *Behavior and Philosophy*, **32**: 97–112.
Velmans, M. (2003), 'Is the World in the Brain, or the Brain in the World?', *Behavioral and Brain Sciences*, **26** (4): 427–9.
Whitehead, A.N. (1967/1925), *Science and the Modern World*, New York, The Free Press.
Zeki, S. (1999), *Inner Vision: An Exploration of Art and the Brain*, Oxford, Oxford University Press.

Lambros Malafouris

The Aesthetics of Material Engagement

The purpose of this chapter is to advance the view of aesthetics as a situated process. The situated process I will be describing here would be neither representational, as for instance in the computational sense found in contemporary neuroaesthetics, nor locational, in the sense of being situated exclusively in any given location (internal or external). Trying to avoid the usual essentialist fallacies, the view of aesthetic experience that this paper builds upon could be described, more simply, as enactive, by which I refer here to an 'open' process of active exploration and material engagement that collapses the boundaries between the mind and the material world. Moreover, rather than identifying aesthetic consciousness with its objects, as in typical analytic philosophy, or use it to define and delimit some pure, detached, and autonomous realm of aesthetic contemplation, the approach taken here aims to ground aesthetic experience into the manifold interfaces of embodied material praxis. The case of pottery making will provide a useful example of such a context of active material engagement and help us to realize a notion of aesthetic experience that is not confined to, or co-extensive with, art or beauty.[1] Moreover this example will enable us also to think about interfaces in terms of mutual permeability and binding rather than separation of brains, bodies and things.

---

1   Understandably this may raise a few problems for the philosopher of art which may prefer the analytical security and closure of a more purified taxonomic approach. For the archaeologist or the anthropologist, however, this contextualization of aesthetic experience constitutes the condition *sine qua non* for its study. Bear in mind, that from an archaeological perspective, conventional pedestals, categories, art-institutions or any other modern framing device are of limited use and applicability in the past for demarcating the prehistory of human aesthetic experience. It is thus important to underline at the outset a key difference in the way archaeology and anthropology approach aesthetic experience relevant to the philosophy of art. This difference is the absence of boundaries. The art object or the aesthetic object is always embedded and functional. It is part of the world and not of an art-world. In archaeology there is no such thing as aesthetic autonomy in the sense of uniqueness and distance from the ordinary world. Aesthetic experience and its objects constitute an integral part of this world.

1 On becoming 'methodological philistines'

From the perspective of archaeology and anthropology the study of aesthetics has always been an extremely difficult subject to tackle. Although space does not permit a detailed review of this field, a basic background of current discussions and debates might be useful. To begin with, three main problems with aesthetics can be pointed out in brief. First, we have the classical problem of the subjectivity of 'aesthetic judgments'. Already from the first archaeological encounters with the expressive material residues of prehistoric 'significant form' (Bell 1958), it became clear that '[a]esthetic judgments are, as the saying goes, matters of taste; and about tastes, as everyone is proud to admit, there is no disputing' (*ibid.*, p. 10). Meanwhile, as the aesthetic debate progressed and matured the value of the intimate association between art experience and aesthetic experience[2] came under increasing critique by many anthropologists studying art and aesthetics in a cross-cultural perspective (Thomas and Pinney 2001; Ingold 1996; Coote and Shelton 1992; Renfrew 2003; Morphy 2007). Finally, more recently, influenced from critical hermeneutic trends (Gadamer 1982; Dewey, 1987; Adorno 1984; Carroll 2001), the focus of analysis shifted away from the aesthetic qualities of the detached object towards the contextual meaning and symbolic content of the intention responsible for its creation. Unfortunately, what was meant to help us understand the cognitive and cultural dynamics of aesthetic experience (response and intent) as a material process, became, more often than not, an anachronistic projection in the past of our own 'aesthetic standards' and modern predispositions about the operation and representational basis of human symbolic thinking (Malafouris 2005; 2007).

It need not be received as a paradox then, that valuable and crucial as it might be for the philosophy of art, within archaeology, the concept of aesthetics started to look epistemologically inappropriate and a hindrance to the interpretation of past material culture and its social and cognitive life. The influential ideas of the anthropologist Alfred Gell offer an excellent point in case.

The methodological stance of *methodological philistinism*, the 'bitter pill very few would be willing to swallow' (p. 161) as Gell characteristically refers in his provocative article *The Technology of Enchantment* (1992), embodies the essence of his thesis about art and aesthetics. The main argument goes somehow like this: unless we are philistines we are forced to 'attribute value to a culturally recognised category of art objects' (*ibid.*, p. 159). However, in our willingness to place ourselves under the spell of art

---

[2] We should not forget that one of the most well rooted beliefs of the Western art tradition is the assumption of a close association between art experience and aesthetic experience, or if you prefer between the art-object and the aesthetic object: something is an artwork if it is designed with the intention to produce/afford aesthetic experience.

we have 'sacralized art' and as such we have become faithful ritual performers where we should have been atheist participant observers. *Methodological philistinism* is being proposed as our only means to remove 'the major stumbling-block in the path of the anthropology of art' and it 'consists of taking an attitude of resolute indifference towards the aesthetic value of works of art' (*ibid.*, p. 161).

For Gell art is not about beauty and taste. Moreover, it is also not primarily about meaning.[3] Only language has meaning as well as the necessary structural consistency for that meaning to be decoded, deciphered and communicated. Objects do not have meanings but social effects. They do not possess semiological qualities to be interpreted, but an agency potential to be abducted.[4] The art-object is not something that represents or encodes symbolic propositions but rather an index of agency and intentionality. As such, our analytic endeavours should concentrate in the examination of the physical qualities, social properties and cognitive mechanisms that make such a thing possible in a given context and at a certain stage of the artefact's biography. For Gell the anthropology of art should not be concerned with aesthetics, but with *'agency, intention, causation, result and transformation'* (1998, p. 6).

The question thus can be raised: do we really need to swallow Gell's 'bitter pill' of 'methodological philistinism'? From what we discussed so far it seems as if the notion of aesthetics, as traditionally defined, has outlived its usefulness. Yet, on the other hand, one could argue that the concept of aesthetics in allowing us to understand the values that people attach to objects in different cultural contexts, or from a long-term archaeological perspective, how humans came to be endowed with the capacity to discriminate between aesthetic and non-aesthetic perceptual forms, objects, and performances, seems too useful to throw out (e.g. Gosden 2001; Morphy 2009).

Is there, then, any way to dissolve this particular incompatibility and save the concept of aesthetics? Answering that, the thing to note first is that this growing depreciation of aesthetics[5] within archaeology and anthropology does not concern, so much, aesthetic experience *per se*, but

3   'I do not deny that works of art are sometimes intended and received as objects of aesthetic appreciation, and that it is sometimes the case that works of art function semiotically, but I specifically reject the notion that they always do' (Gell 1998, p. 66).

4   Indeed, Gell adopts the term of abduction as the proper designation for the cognitive operation we 'bring to bear on indexes'. We should note that Gell being concerned with indexes that 'permit the abduction of "agency" and specifically social agency' uses this term in order to designate 'a class of semiotic inferences which are, by definition, wholly distinct from the semiotic inferences we bring to bear on the understanding of language' (Gell 1998, p. 15).

5   Naturally, this growing depreciation did not stand in isolation from similar critical reactions into the field of philosophy. However, although contemporary trends and tensions within the philosophy of art certainly have contributed in making the concept of aesthetics increasingly suspicious, the anthropological debate over art and aesthetics has grown out of its own peculiar intellectual environment.

rather, the analytical value of aesthetic experience as an object of anthropological study. For instance, when Gell rejects aesthetics he is referring to the peculiar Western understanding of beauty and the inherent subjectivity and biased character of this enterprise:

> I am far from convinced that every 'culture' has a component of its ideational system which is comparable to our own 'aesthetics'. I think that the desire to see art of other cultures aesthetically tells us more about our own ideology and its quasi-religious veneration of art objects as aesthetic talismans, than it does about these other cultures. (Gell 1998, p. 3)

Indeed, reformulating in his *Art and Agency* (1998) the question of art from that of *'what is art?'* to that of *'what does art do?'*, Gell introduces agency as the new measure of art-like situations. In that sense, asking us to become detached from our aesthetic predispositions in order to discover the agency of the artwork was indeed a breakthrough of immense potential. Yet asking us to erase the word aesthetic altogether from our vocabulary is something else. It is indeed the case that our aesthetic presumptions, in the common sense of taste and beauty, can be a great obstacle when examining the aesthetic agency of things or artworks. But this does not mean that aesthetics should be abandoned as a category, it means that aesthetics should be placed on a new ontological basis. Similarly, that aesthetic theories of art have failed in many important respects does not necessarily mean that aesthetic experience is a useless term; it means that those theories can no longer be treated as the grounding framework for the anthropology and archaeology of art.

I believe Gell's mistake was to turn what is a valid methodological suggestion into an invalid eliminative argument that misconstrues the problem with aesthetics. Adopting the attitude of aesthetic detachment is not problematic for an anthropology or archaeology of art due to the inherent subjectivism and ethnocentrism that such a stance embodies, but because it leads us to differentiate between domains of experience that were probably inextricably enfolded. To exemplify, while in the context of Western aesthetic theory a thing can have agency only as an aesthetic object—that is, as an object designed with the intention to produce/afford aesthetic experience—in the context of prehistoric and ethnographic art the arrow of this relationship points in both directions. That means that a thing can have agency as an aesthetic object, but also *a thing can become an aesthetic object because it has agency*.[6] The issue is not to deny the first for the sake of the second but to recognize both, that is to recognize the *aesthetic of agency*.

What we need then, is a situated aesthetic approach that will give us the opportunity to explore alternative forms of aesthetic experience and ways

---

6 For instance, robots and other forms of AL and AI.

of seeing. We need to place aesthetics against a more appropriate ontological foundation not to reject the whole enterprise. The latter would deprive of us the ability to develop an understanding 'of how other cultures "see" the world' (Coote and Shelton 1992, p. 8) and understand what it means to make things. Following that, to the degree that I will become a sort of 'methodological philistine' in the remaining of this paper, my intention is not to denounce aesthetics for the sake of action and material agency but instead to ground aesthetics in situated embodied action.

2 At the potter's wheel

Imagine a potter throwing a vase on the wheel (Malafouris 2008a). Try to follow the complex orchestration of action as this unfolds throughout the different stages of this creative process. What sort of projections, relations, or representations are needed for this aesthetic performance of active material engagement by which an amorphous mass of plastic clay is transformed into something we might call an 'aesthetic object'?

Traditionally, to answer a question like that one would have to choose between two major approaches: one approach is to look beyond skin, at the final end of any given artistic process, for the properties of aesthetic objects. On the analytical level, aesthetic objects have the advantage of being potentially well defined and sufficiently static. They can be perceived with sufficient clarity, named, classified and talked about, while different aesthetic characteristics can be attributed to them. On the phenomenological and anthropological level, however, they fail to accommodate the temporal dynamics and embodied aspects of aesthetic experience. For, what is an aesthetic object really? Even philosophers like Monroe Beardsley with a key role in the formulation of aesthetic theory, were very much aware of the many problems that surround the use of such a term: in what sense can we speak in the context of aesthetics of an 'object'? Looking at the biography or life history of an art work 'it appears as a more or less continuous process, in which shifts and transformations are constantly going on... Now, where, you might ask, is there an *object* in all this? Does anything stand still long enough to be called an "object"?' (Beardsley 1981, pp. 16–7).

Turning now to the second major approach to answer our question about the aesthetics of pottery-making this would be to look at the other end of the process, trying to account, by looking inside the mind/brain, for the origins of aesthetic experience. Aesthetic experience now becomes a quality of mind rather than of objects or things. It exists inside the subjects's head as an affective or intentional mental state. Consider for instance the moment where the potter's intention to act is formed. What is it that guides the dextrous positioning of the potter's hands and decides upon the precise amount of forward or downward pressure necessary for centring a lump of clay on the wheel? How do the potter's fingers come to know the precise force, place and time of the appropriate grip? What kind of cognitive recourses (neural and extra-neural) are involved? From a neuroscience perspective the obvious way to deal with the inextricable dynamic coupling between the potter's fingers, body and task-environment is to erroneously assume (for a more detailed argument see Malafouris 2008a, also discussion of neuroaesthetics in the final section) that the potter's fingers and body do nothing but transmit information to and execute the orders of the potter's brain. However, in the real context of material engagement and situated action there is much more than mere motor control processes that constitute the effortless movement of the potter's hand as it reaches for, and gradually shapes, the wet clay. First the hand grasps the clay in the way the clay affords to be grasped, then the action becomes skill, skill effects results and from those results that matter aesthetic agency emerges. In fact, a whole set of conceptual challenges to the embodied nature of action and the meaning of aesthetic agency, object and intent can be raised (see Malafouris 2008a).

Someone may object here, that our previous points have very little to do with aesthetic experience as such. Instead, they relate primarily to the issue of embodiment, sensorimotor control, tacit knowledge and practical skill and thus have very little to tell us about aesthetics proper. But this is precisely the assumption that I want to question by collapsing the dividing lines between perception, cognition and action, and rejecting the methodological separation between aesthetic experience and embodiment. How can this be done? I suggest there are two basic obstacles that one needs to overcome. First we need to rethink our conventional philosophical understandings of aesthetic experience in terms of isolated objects, products, or intentions and their concomitant 'passive recipient' model of aesthetic reception. For instance, we usually speak of aesthetic objects or experiences and their properties in a 'passive' tense without paying proper attention to the constructive process and materiality of aesthetic creation as a form of action. As a consequence aesthetics very often looks more like a 'happening' than a 'doing'; something suffered or undergone by a 'patient' rather than actively engaged in. This is also why many people may find difficult to conceive pottery making as an 'aesthetic performance'. Trying to sketch a different conception of aesthetic experience as a situated process our first step should be to break away from this dominant trend.

This brings us to the second obstacle that we need to overcome and which concerns the choice of the right boundaries for our unit of analysis. This of course relates to another more basic question that has recently attracted a great deal of attention in philosophy of embodied and situated cognition: where do we draw a line between mind and the material world? How can we best understand the nature, and infer the direction, of the causal links between brains, bodies and things? One way to answer that can be found by extending and expanding the conventional demarcation lines of skin and skull into the world (e.g. Clark 1997; Clark and Chalmers 1998). Another way might be to abandon once and for all the logic of 'boundaries' and 'delimiting lines'. Indeed, do we really need to draw a line? Even to think in terms of separating lines would be to commit the usual Cartesian fallacies. Lines when used as boundaries or epistemological constrains can help us re-shape and rethink a pre-established relationship between different, and quite often incommensurable, ontological domains but they cannot provide the ontological unity we are after. The purity of delimiting lines or of any concomitant neat analytical or metaphysical distinction cannot accommodate the transgressing power of embodied and mediated action. Consider, for instance in our example of pottery making, the intriguing blending and binding of the various ontological ingredients of action (physical, bodily, mental, artefactual or biological). Where does the potter's brain and body stop and the world begin?

For all these reasons, it can be argued that the sort of 'no-line' approach serves better the epistemic demands of situated action and material engagement. Nonetheless, it is also the case that many times the key problem facing us is not about *where* or *if* we should draw a line, but instead about the *kind* of line that we choose to draw.

The insights offered in the latest work of the anthropologist Tim Ingold on the ontology of lines (2007; 2008) might be of some use in this context.

3 What kind of line? 'A view from the open'

In the beginning of his thought provoking paper, 'Bindings Against Boundaries', Ingold invites us to take a pen and draw a rough circle on a sheet of paper (Ingold 2008, p. 1796). What do we see when we look at this shape? How can we interpret and perceive this line we have just produced? Ingold suggests that there are two ways: the first way is to see the line as the end or final product of human intention and design abilities. The line then becomes a static geometric perimeter that delineates the form of a circle, that is, a fixed totality. But there is a second way of seeing the line. Instead of a fixed totality we could see in this line the dynamic trace of human gesture. The line then becomes the index of an open process. It resembles a vector more than a static shape; the trajectory of a movement more than the perimeter of a figure.

According to Ingold, each of the above ways of seeing emanates from, reiterates, and signifies the operation of a specific logic which determines our ways of thinking about human design, and by extension, I may also suggest, of human aesthetic experience. In particular, the logic operating behind the seeing of a circle refers to a tendency, deeply entrenched in the structure of modern thinking, to 'turn the pathways along which life is lived into boundaries within which life is contained'. Tim Ingold calls it the 'logic of inversion' and suggests that it is responsible for transforming our perception of the line from that of an active trajectory of movement, to a dividing line between 'what is on the "inside" and what is on the "outside"' (*ibid.*, p. 1796). Put now this 'logic of inversion' into reverse and you get the second kind of logic, one that sees lines and force trajectories instead of boundaries and closed circles. This is the 'logic of inhabitation' and it is about an 'open' world, i.e. a world that is inhabited rather than occupied:

> In the open world, the task of habitation is to bind substances and the medium into living forms. But bindings are not boundaries, and they no more contain the world, or enclose it, than does a knot contain the threads from which it is tied. To inhabit the open is not, then, to be stranded on a closed surface but to be immersed in the fluxes of the medium. (*ibid.*)

It is precisely this shift from 'occupation' to 'inhabitation' that I find particularly useful for our current discussion of situated aesthetics.[7] Taking inspiration from Ingold's suggestion I want to propose that the world of aesthetics is inhabited rather than occupied, it is not furnished with already-existing things but instead woven from the strands of their continual coming-into-being. This has some very important implications in the way aesthetic experience should be approached: simply put, instead of static aesthetic 'objects' we should be seeking for active aesthetic 'processes' that grow and live 'in the open'. Of course, to be 'in the open' embodies something of an oxymoron. It simultaneously signifies the placements within limits—'in'—and the absence of limit—'openness' (Ingold 2008, p. 1797). The question then becomes, as we argued in the previous section, not *if or where* to draw a line but rather *what* sort of line should we draw? Trying to answer that, I suggest we should abandon the usual 'closed' or 'delimiting' type of lines aiming at demarcating and isolating domains of experience (e.g. mental and physical, biological and cultural, etc.) and replace them with 'connecting' lines of active material engagement. Lines of this sort are no longer solid boundaries but dynamic interfaces and 'surrogate structures' (Clark 2010) that can help us visualize the invisible bindings and trajectories of human action.

4 Enactive discovery

It follows from our previous discussion that our attempt to construct a 'situated aesthetic' presupposes, on the one hand, avoiding the representational fallacies, and on the other, finding a way of turning the inanimate

---

7 As Ingold points out here this relates to what the philosopher Martin Heidegger (1971) identified as the foundational sense of dwelling: not the occupation of a world already built but the very process of inhabiting the earth.

stuff of modern 'occupation' design in to something alive and 'inhabitable'. How this can be done? What could be our escape route from the representational predicament of conventional 'aestheticism'? One possible strategy would be to follow a thread of thought that initiates in the rhizomatics of Deleuze and Guattari (1987), passes through Ingold's conception of life as an 'ever-ramifying bundle of lines of growth' (2008, p. 1807), and finally ends up at an idea of a 'meshwork' (Varela 1991; de Landa n.d.; Ingold 2007). Another, and I suggest more simple, solution is to adopt the perspective of the material engagement approach (Malafouris 2004; Renfrew 2003) and follow the enactive thread that it embodies. Choosing material engagement as our escape route essentially means that we ground our approach in a specific sense of embodiment which we may call *active material engagement*. The differentiating feature of active embodiment is that it focuses on the *act of embodying* rather than on the conventional sense of embodiment as passive incorporation (see Malafouris 2008b). I suggest that this type of decentralized or distributed intelligence, what we may call *enactive discovery*, should occupy centre stage in any conception of situated aesthetics.

The choice of the term, i.e. *enactive discovery*, besides the obvious link with the so-called enactive paradigm in philosophy of mind (e.g. Noë 2009; 2005), signifies also the intellectual kinship of the view of situated aesthetics advanced here with some aspects of traditional aesthetic theory, seen here in my rather selective use of Monroe Beardsley's notion of 'active discovery' (Beardsley 1982). I should clarify that drawing on the notion of 'active discovery' my purpose is not to subscribe to Beardsley's aesthetic point of view but simply to embrace and re-animate something of the traditional vitality of this notion especially in the face of the growing marginalization of aesthetic experience both in archaeology and anthropology (e.g. Gell 1998; see previous section) as well as in analytic philosophy (see, e.g. Dickie 1974; 1984; Danto 1964; 1981; 1986; 1997; Goodman 1968; Dewey 1987). It should be noted that in the context of Beardsley's theory 'active discovery' (1982, p. 188–9) refers to the quality or property of intelligibility that accompanies the discovery of connections between percepts and meanings. This sense or quality is, according to Beardsley, one of the five central requirement of aesthetic experience — the other four being object directness, felt freedom, detached effect, and wholeness (*ibid.*). My own formulation of this term as 'enactive discovery', although it retains some of the initial 'externalist' connotations, now refers primarily to connections that transcend the immediacy of the aesthetic object as a fixed end-product. Enactive discovery, instead, refers to the totality of situated praxis as a trajectory of material engagement.[8] This unfolding trajec-

---

8   Another way to see enactive discovery is as a process of 'we' or 'joint' intentionality (Tomasello *et al.* 2005) that is now extended beyond the sphere of human interaction to include the material

tory becomes, for the potter, simultaneously the tool and product of an active exploration.

To illustrate that better a useful comparison can be drawn here with the work of the painter Jackson Pollock. A similar shift in perspective from static objects to active situated processes can be seen realized in the way Pollock has radically changed the conventional positioning of the canvas (from vertical to horizontal) and the relationship of the body to the painting. His action-painting has efficiently turned attention away from the fixed final object (the canvas) to the act of painting itself. The final 'aesthetic' object that we see on the canvas is turned now into an index of action inseparable from Pollock's gesturing body. Object and process, body and line, perception and action, all the ingredients of action (mental and physical) are now inextricably intertwined into an aesthetic biographic compound.

5 The affect of engagement and the aesthetics of agency

We spoke above of 'enactive discovery' and suggested, using the example of pottery making, that aesthetic experience is something that we *do* rather than something that *happens to* us, or *in* us (see, e.g. Noë 2001; 2005). But what is it more precisely that the above formulation implies for the study of aesthetics? Though enactive discovery by no means provides a necessary or sufficient condition for aesthetic experience, we suggested it might be regarded as a background condition for the study of aesthetics. In what ways then might this notion help us rethink the differentiating, evaluative and phenomenological dimensions of aesthetic experience?

There are many directions that one could follow in order to explore these questions. Here I will focus on and briefly discuss two central notions, i.e. agency and perception.

Agency and intentionality

As the sociologists J. Law and A. Mol (2008, p. 57) observe, 'questions about agency, are usually asked as part of a search for explanation. What is the origin of an event?' In our case, the event we are seeking to understand concerns the aesthetic dimensions of pottery making. Thus, it may seem initially that before we speak of aesthetic agency we need, at a first level, to construct a temporal and causal hierarchy of action. This will enable us, at a second level, to separate and prioritize among the various ontological ingredients of action (e.g. intentions, processes, techniques, objects and materials). The problem with traditional approaches to aesthetic experience is that, as discussed in a previous section, they either focus at the final product of the chain of action looking for the aesthetic qualities in things

world. In this sense it refers to the irreducible dance between human and material agency through which the 'affordances' of materials are discovered.

themselves, or they would focus inside the subject's head for the origins of aesthetic intent and/or experience. The first tendency assigns primacy to the properties of the final product, i.e. the aesthetic object. The second tendency assigns primacy to the 'internal' processes of formation, i.e. the aesthetic intent, as against their final products. Yet in both cases the actual flow of action as it emerges through the embedded concerted activity of the whole situated person is lost. Is there a way out of this?

The view of aesthetic experience as a situated process proposed here aims precisely at offering an escape route from that problem by collapsing and criss-crossing those artificial boundaries. The model of situated aesthetics proposed here assigns primacy to the transformative processes of material engagement. Which is to say it ascribes primacy to none of the elements (internal or external), but instead, to the very linkages and bindings that transform the amorphous mass of clay into an aesthetic object.

It is important to understand that the problem with aesthetic agency in mediated action is that the purity of action has been lost. The potter's body, being messy and leaky (Clark 1997), cannot be as rigidly defined and circumscribed as traditional aesthetic theories might prefer. The potter's body, enmeshed into the realm of skill, learning and practice we call pottery making, becomes a different body. The *being* of the potter is co-dependent and interweaved with the *becoming* of the pot. Constantly extending and incorporating the non-biological it becomes *more than a body; a situated body*.[9]

This also means that the constituents of aesthetic experience are not to be found before or outside the throwing or the shaping of the pot. The constituents of aesthetic experience are *in the throwing, in the shaping*. They can be understood as a mixture or blend of ingredients that can act or be acted upon. Trying to separate brain, body, and material culture in the above nexus of active material engagement 'is like trying to construct a pot keeping your hands clean from the mud' (Malafouris 2008a). Aesthetic experience becomes then a binding of materials; a dynamic flow of the organic into the inorganic. It is in this sense that which I call *enactive discovery* should be understood.

This brings us to the issue of perception.

Enactive perception

What does the potter's experience of seeing and touching the clay really consist of? Take, for instance, the question about the visual prototypes and perceptual categories that motivate the shape and final form of the clay

---

[9] However, 'being situated' does not simply mean that the potter is located somewhere for the same reason that 'being embodied' does not simply mean that the potter is having a body. Instead, 'being situated' means that 'situatedness' *matters*, it means in other words that the situation (environmental, technological, historical or social) can shape and/or become part of the actual aesthetic process or experience.

vase to be produced. Where are these to be found? How do they operate in action? A neuroscientist, for instance, following the inherently reductive and neurocentric logic that characterizes much of contemporary neuroaesthetics will attempt to locate the neural substrates or correlates of these processes inside the brain. For example, in the case of the visual prototypes, the area of the inferior temporal cortex (ITC) might be a good bet given that category-specific agnosias follow its damage (Palmeri and Gauthier 2004). But what does this strategy of 'localization' really tell us about aesthetic perception? I suggest, not much. For one thing, I believe that the underlying 'internalist' assumptions in neuroaesthetics that one sees with the brain (see, e.g. Zeki 1998; 1999; 2002), or that there might be inside our brains some universal neural signature of beauty (e.g. Kawabata and Zeki 2004), are deeply misconceived. For another, and more relevant to our own example of the potter, vision is only one modality of the cross-modal and synaesthetic ensemble active in pottery manufacture. That means that the search for neural correlates of visual categories in the inferior temporal cortex (ITC), or in any other brain region critical for visual recognition, or even at some point along the retina-brain computational circuitry, will not help us understand much about the actual phenomenon of interest. Let me clarify that what I deny here is not the existence of 'neural correlates' of aesthetic experience, perception or beauty. I see no reason to disagree, for instance, that 'judging a painting as beautiful or not correlates with specific brain structures, principally the orbito-frontal cortex, known to be engaged during the perception of rewarding stimuli and, perhaps surprisingly, the motor cortex' (*ibid.*, p. 1702). So far as I can tell there might well be — or potentially constructed — a neural correlate, substrate, or representation of anything in the sphere of human experience. No doubt, human aesthetic experience, in all its different manifestations, must have some sort of material basis or grounding in the brain. What, however, I categorically reject is that such a 'neural correlate' can be seen itself as 'sufficient', or as having any 'ontological' priority or 'constitutive' role in the production of human aesthetic experience (including but not limited to beauty).[10] With this statement I do not simply want to point out that external factors influence the shape of the 'internal representations' that support the phenomenal or intentional states responsible for the production of aesthetic experience, but instead, something more radical, namely, that varieties of perceptual or aesthetic experience can emerge in the absence of any corresponding universal 'internal representations' or fixed neural underpinnings to support them. In other words, when it comes to aesthetic experience, similarly to many other cognitive

---

10 For a similar critical argument about the role of the mirror neuron system as providing the neural basis of empathic response to artworks see (Casati and Pignocchi 2007).

and affective phenomena, brain is only part of the story.[11] Or as Noë puts it, and as our example of pottery making exemplifies, 'we are out of our heads' (Noë 2009, p. xiii). As a result, useful as it might be in the context of other problems and hypotheses, the 'neural correlate' or 'signature' approach has little to offer in the understanding of aesthetic experience. To understand aesthetic experience in its true material dimensions we need to avoid all different sorts of unwarranted reductions and adopt an integrative enactive framework of thinking.

A comparison — somewhat simplistic — of the potter's experience with the experience of the feeling of a bottle discussed by the philosopher Alva Noë (2001, pp. 48–51; 2005) may help to clarify these points. Imagine that you hold a bottle with your eyes shut. Naturally you do not hold the whole bottle in your hand but only those parts of the bottle's surface at the points where the bottle touches your skin. Nevertheless, the feeling or experience you get is not that of holding a partial bottle but rather that of perceiving the totality of the bottle. In other words it seems as if the whole bottle, and not just one's finger-to-bottle points of contact, is present to your awareness. Why is that? One way to answer that question, following the traditional 'internalist' suppositions, would be that our awareness of the bottle is based on the availability of an internal representation about this bottle. Our brain produces this internal model on the basis of the partial information made available by touching the bottle's surface. It is the presence — inside our heads — of such a model about the bottle which enables us to have the experience of the bottle as a totality. Another way to answer our question would be to deny the validity of our characterization of what the experience is like. Contrary to the way we seem to experience the untouched parts of the bottle, such an experience of sensing the bottle as a whole is essentially a confabulation. There is, nonetheless, a third way to account for the experience of the bottle we take ourselves to have, namely, the enactive account. As Noë suggests we do not need either to deny ourselves the experience of being aware of the bottle and sense it as a whole, or go as far as saying that we truly are in contact with all parts of the bottle's surface. Rather, it seems 'as if the bottle, as a whole, is present' and that we have 'access to detail about currently untouched parts of the bottle by further hand movements' (Noë 2001, p. 49). According to Noë the same point can be extended in the case of vision:

> When we see, it perceptually seems to us as if we encounter a world of immense detail, but it doesn't seem to us as if we have all that environmental detail in our minds at once, no more than it seems to us as if we make contact with all parts of the bottle's surface. Rather, it seems to us as if the environment is there, and as if we can extract information

---

11 In fact the only perspective point from which art, as S. Zeki seems to suggest (Zeki 1998, p. 78), can be seen as a true extension or prosthesis of the brain is that of a mind not limited by the skin.

from the environment by moving our eyes or head or by reorienting our bodies. (*ibid.*, p. 49)

It is here that the connection with our previous discussion of aesthetics become more obvious. How does the potter know to make a beautiful vase?

6 The enactive logic of pottery-making as an aesthetic process

Drawing on the discussed enactive paradigm one possible answer to the above question of how to account for the potter's aesthetic perception and experience can be found by way of the so-called 'sensorimotor contingencies' (O'Regan and Noë 2001). These 'sensorimotor contigencies' enable the hand of the potter to navigate upon the surface of clay with a minimal need of internal processing, using, instead, the world as an outside memory (see O'Regan 1992; see also Brooks 1991). All resources needed are right there; available at hand within reach. In other words, the materiality of the 'situation' into which the potter is embedded becomes a vehicle of outside memory through which the potter's imagination and creativity acquire also non-biological extra-neural properties. Material engagement offers an interface where neural plasticity and the plasticity of clay meet (Malafouris 2008a; 2010).

Accepting that may sound initially confusing but is in fact the opposite. The reason is simple: dwelling in this continuum of active engagement saves us the trouble of having to look inside the potter's head for the putative traces of aesthetic experience. The traces of aesthetic experience exist out there, at the potter's wheel, and become available to us via the empirical signatures left by the 'sensorimotor contigencies' of the act. It is the skill-based confidence on the part of the potter as a perceiver that s/he is able to acquire the information needed through simple bodily movement (via sight, touch, or other perceptual modality), that forms the basis for the feeling of the perceptual contact and presence of the surrounding environment as a whole.

The potter's perceptually-guided exploration of the surface of the clay vessel as it gradually emerges and takes shapes does not require any internalized pre-formed model about the final shape of the vessel to be produced.

The potter's sensory apparatus is not a passive means that allows the transduction of the necessary information inside the potter's head where it can be subsequently manipulated. The potter's sensing should be better perceived as a dynamic interface of engaging the world which constantly shifts location. Sometimes it is sensed where the hand meets the surface of the clay and sometimes where clay meets the potter's eye.

In other words, it is the totality of the relevant material environment rather than some preformed internal representation of it inside the potter's head that is being engaged during the unfolding of the creative process. The final form of this shape is to be decided *in action*, as the emergent product of the potter's movement and skillful active material engagement. When this process of active material engagement is accompanied by the presence of 'enactive discovery' it produces the basis for 'aesthetic experience'. The transient and temporary character of these dynamic ensembles should not lead us to downgrade either their cognitive or aesthetic status.

References

Adorno, T.W. (1984), *Aesthetic Theory*, New York, Routledge & Kegan Paul.
Bell, C. (1958), *Art*, New York, Capricorn Books.
Beardsley, M.C. (1981), *Aesthetics: Problems in the Philosophy of Criticism*, New York, Hlreourt. (Original work published 1958.)
Beardsley, M.C. (1982), *The Aesthetic Point of View*, Ithaca (NY), Cornell University Press.
Brooks, R.A. (1991), 'Intelligence Without Representation', *Artificial Intelligence*, **47**: 139-59.
Carroll, N. (2001), *Beyond Aesthetics: Philosophical Essays*, Cambridge, Cambridge University Press.
Casati, R. and A. Pignocchi (2007), 'Mirror and Canonical Neurons are not Constitutive of Aesthetic Responses', *Trends in Cognitive Sciences*, **11**: 410.
Clark, A. (1997), *Being There: Putting Brain, Body and World Together Again*, Cambridge (MA), MIT Press.
Clark, A. (2010), 'Material Suurogacy and the Supernatural' in L. Malafouris and C. Renfrew, Eds., *The Cognitive Life of Things: Recasting the Boundaries of the Mind*, Cambridge, McDonald Institute Monographs: 23-8.
Clark, A. and D. Chalmers (1998), 'The Extended Mind', *Analysis*, **58** (1): 10-23.
Coote, J. and A. Shelton (1992), 'Introduction' in J. Coote and A. Shelton, Eds., *Anthropology, Art and Aesthetics*, Oxford, Clarendon Press: 1-11.
Danto, A. (1964), 'The Artworld', *Journal of Philosophy*, October 15: 571-84.
Danto, A. (1981), *The Transfiguration of the Commonplace: A Philosophy of Art*, Harvard, Harvard University Press.
Danto, A. (1986), *The Philosophical Disenfranchisement of Art*, New York, Columbia University Press.
Danto, A. (1997), *After the End of Art: Contemporary Art and the Pale of History*, Ithaca (NY), Princeton University Press.
de Landa, M. (n. d.), *Meshoworks, Hierarchies and Interfaces*, [Online], http://www.t0.or.at/delanda/
Deleuze, G. and F. Guattari (1987), *A Thousand Plateaus*, Minneapolis, University of Minnesota Press.
Dewey, J. (1987), *Art as Experience*, Carbondale (IL), Southern Illinois University Press.
Dickie, G. (1974), *Art and the Aesthetic: An Institutional Analysis*, Ithaca (NY), Cornell University Press.
Dickie, G. (1984), *The Art Circle*, New York, Haven.
Gadamer, H.G. (1982), *Truth and Method*, New York, Crossroad.
Gell, A. (1992), 'The Technology of Enchantment and the Enchantment of Technology' in J. Coote and A. Shelton, Eds., *Anthropology, Art and Aesthetics*, Oxford, Clarendon Press: 40-63.
Gell, A. (1998), *Art and Agency: An Anthropological Theory*, Oxford, Oxford University Press.

Gosden, C. (2001), 'Making Sense: Archaeology and Aesthetics', *World Archaeology*, **33** (2): 163-7.
Goodman, N. (1968), *Languages of Art*, Oxford, Oxford University Press.
Heidegger, M. (1971), *Poetry, Language, Thought*, A. Hofstadter, Trans., New York, Harper and Row.
Ingold, T., Ed. (1996), *Key Debates in Anthropology* ('Aesthetics is a Cross-Cultural Category'), London, Routledge.
Ingold, T. (2007), *Lines: A Brief History*, London, Routledge.
Ingold, T. (2008), 'Bindings Against Boundaries: Entanglements of Life in an Open World', *Environment and Planning*, **A** (40): 1796-810.
Kawabata, H. and S. Zeki (2004), 'Neural Correlates of Beauty', *Journal of Neurophysiology*, **91**: 1699-705.
Law, J. and A. Mol (2008), 'The Actor-Enacted: Cumbrian Sheep in 2001' in C. Knappett and L. Malafouris, Eds., *Material Agency: Towards a Non-Anthropocentric Perspective*, New York, Springer: 57-78.
Malafouris, L. (2004), 'The Cognitive Basis of Material Engagement: Where Brain, Body and Culture Conflate' in E. DeMarrais, C. Gosden and C. Renfrew, Eds., *Rethinking Materiality: The Engagement of Mind with the Material World*, Cambridge, The McDonald Institute for Archaeological Research: 53-62.
Malafouris, L. (2005), *Projections in Matter: Material Engagement and the Mycenaean Becoming*, Unpublished PhD dissertation, Cambridge University.
Malafouris, L. (2007), 'Before and Beyond Representation: Towards an Enactive Conception of the Palaeolithic Image' in C. Renfrew and I. Morley, Eds., *Image and Imagination: A Global History of Figurative Representation*, Cambridge, McDonald Institute for Archaeological Research: 289-302.
Malafouris, L. (2008a), 'At the Potter's Wheel: An Argument for Material Agency' in C. Knappett and L. Malafouris, Eds., *Material Agency: Towards a Non-Anthropocentric Perspective*, New York, Springer: 19-36.
Malafouris, L. (2008b), 'Beads for a Plastic Mind: The "Blind Man's Stick" (BMS) Hypothesis and the Active Nature of Material Culture', *Cambridge Archaeological Journal*, **18** (3): 401-14.
Malafouris, L. (2010), 'Metaplasticity and the Human Becoming: Principles of Neuroarchaeology', *Journal of Anthropological Sciences*, **88**: 49-72.
Morphy, H. (2007), *Becoming Art: Exploring Cross-Cultural Categories*, Oxford, Berg.
Morphy, H. (2009), 'Art as a Mode of Action: Some Problems with Gell's Art and Agency', *Journal of Material Culture*, **14** (1): 5-27.
Noë, A. (2001), 'Experience and the Active Mind', *Synthese*, **129**: 41-60.
Noë, A. (2005), *Action in Perception*, Cambridge (MA), MIT Press.
Noë, A. (2009), *Out of Our Heads*, London, Macmillan.
O'Regan, J.K. (1992), 'Solving the "Real" Mysteries of Visual Perception: The World as an Outside Memory', *Canadian Journal of Psychology*, **46** (3): 461-88.
O'Regan, J.K. and A. Noë (2001), 'A Sensorimotor Approach to Vision and Visual Perception', *Behavioral and Brain Sciences*, **24** (5): 939-73.
Palmeri, T. and I. Gauthier (2004), 'Visual Object Understanding', *Nature Reviews Neuroscience*, **5**: 291-303.
Renfrew, C. (2003), *Figuring It Out: The Parallel Visions of Artists and Archaeologists*, London, Thames & Hudson.
Thomas, N. and C. Pinney, Eds. (2001), *Beyond Aesthetics: Art and the Technologies of Enchantment*, Oxford, Berg.
Tomasello, M., M. Carpenter, J. Call, T. Behne and H. Moll (2005), 'Understanding and Sharing Intentions: The Origins of Cultural Cognition', *Behavioral and Brain Sciences*, **28**: 675-735.
Varela, F. (1991), 'Organism: A Meshwork of Selfless Selves' in A.I. Tauber, Ed., *Organism and the Origins of Self*, The Hague, Kluwer: 79-107.

Zeki, S. (1998), 'Art and the Brain', *Daedalus: Proceedings of the American Academy of Arts and Sciences*, **127** (2): 71–104.

Zeki, S. (1999), *Inner Vision: An Exploration of Art and the Brain*, Oxford, Oxford University Press.

Zeki, S. (2002), 'Neural Concept Formation and Art: Dante, Michelangelo, Wagner', *Journal of Consciousness Studies*, **9** (3): 53–76.

Sabine Marienberg

Language, Rhythm, Grain of Voice

While ordinary language predicators are normally learned implicitly and empractically, i.e. by increasingly understanding how they are used in a linguistic community, scientific and philosophical terms are not. They are explicitly introduced with definitions or examples and counterexamples. Unfortunately, defining 'situated aesthetics' as aesthetics which are situated in a certain way would not only be trivial, but would also inevitably lead to a series of follow-up questions: is 'aesthetics' supposed to refer to sensual perception in general, or to the reception and production of works of art? And if the latter, to measurable sense-physiological responses to them, or to aesthetics as a theory of the so-called lower cognitive faculties, or to questions of taste? And how should we understand 'situated' — does it mean localizable in a certain place, as for instance in England or Greece? In a specific time or a well-delimited cultural setting (which would at least do some justice to the plural form 'aesthetics')? Or, as the subtitle of this volume, 'beyond the skin', suggests, as both inside and outside the human body (which would mean everywhere)?

As one can easily see, we already find ourselves in the midst of a somewhat erratic search for explicit examples—which is not a big surprise, given that definitions, without being linked to experiences, are either axiomatic or empty. If we try to look for counterexamples, we are no better off: for just aesthetics we could at least point to merely mental operations or imperceptible objects and phenomena (like molecules or radioactivity), but what on earth would non-situated aesthetics be? Aesthetics independent of sensing subjects and/or sensible objects? Ubiquitous, time-transcending, resistant to taste?

Without expanding the list of possible examples and counterexamples any further, we could focus on a special case of aesthetic experience, namely the various forms of experiencing language—and investigate to what extent these could be seen as situated, and how the notion of situated aesthetics could help us to understand the processes involved. The aim in the following is twofold: to heuristically differentiate the concept of situatedness by means of an actual example, and to figure out whether it allows us to discover something so far unknown about linguistic experience. In other words 'situated aesthetics' will be taken as a starting point to explore certain traits of language production and perception, with particular attention to poetic speech. The point is not to attribute aesthetic value through artistic evaluation or judgments of taste, but rather to explore what language in its artistic aspects reveals about what we do and suffer in experiencing language in general.

1 Experience

To experience something *as* something, a person has to perceive it physically, but it also must be practically or theoretically articulated — since in order to be treated or conceived in a certain way, it must be distinguishable from other things, be it from other objects/events or from the subject itself.

Whether something can be experienced at all depends on a subject's capacity for sensing (exposure to music would be meaningless to a deaf person), a capacity to feel certain qualities in isolation, still unconnected to anything else: the possibility of blue, of something warm or soft. Yet what constitutes an experience is the replacement of one feeling by another. This is what Charles Sanders Peirce called the *sense of reaction*, and it manifests itself as surprise or disappointment, a forcible disruption between what was before and what is now, or what was expected and what was found. It is, in other words, the notion of an antagonism, not only between two feelings, but also between ourselves and what confronts us.[1] Thus to be experienced in something means to have it learned the hard way, in the entwinement of activity and passivity (see also Schneider 1992). But in order to be experienced *as* something, say a man as a painter (and not as a husband or as someone who just caught a cold) or a written word as a linguistic entity (and not as a meaningless accumulation of ink on paper), it has to be conceptualized. If we don't want to live in an ever-changing world of threats and wonders, we had better be able to find explanations that allow us to let surprises seem less surprising and to integrate errors and disappointments into a learning process.

Imagine someone lying in a meadow with his eyes closed, enjoying the warmth of the sun on his skin and not thinking of anything in particular, when all of a sudden he feels something cold and wet all over. His situa-

---

[1] 'This sense of acting and of being acted upon, which is our sense of the reality of things, — both of outward things and of ourselves, — may be called the sense of Reaction. It does not reside in any one Feeling; it comes upon the breaking of one feeling by another feeling. It essentially involves two things acting upon one another' (Peirce 1998, p. 6).

tion will change dramatically, and as a first reaction he might jump up furiously. But when he starts to look around in bewilderment and spots a sprinkler that escaped his notice before, his astonishment will give way to insight. And not only will he learn that the sprinkler exists (and that sprinklers do sprinkle, in case he had never seen one before), but also that he himself hadn't expected it to be there. Of course someone could tell him in advance, but this would only work if he knew from experience at least some of the elements involved. If he didn't have first-hand knowledge of sprinklers or the feeling of getting wet, telling him that if he stayed where he was he'd soon be pretty drenched would be the same as telling him that 'when the nearby swawler started graunching, he'd get emelded'. Knowing everything only from hearsay is as pointless as identifying the position of a place we are looking for between two other equally unknown positions. Yet feeling alone or the ability to sense action and reaction without the ability to identify invariants and general rules in the world would deprive us of any possible orientation. But what, after all, would be a world without surprises?

With this in mind, we could propose to interpret 'situatedness' as characterizing an aspect of experience. An experience may simply happen to someone, or he can at the same time know about it. In the latter case, i.e. in a moment in which something 'takes shape' or 'makes sense' for someone due to an experience, we have a reflection upon situatedness (which can also be reflected on in turn).

Getting back to language, we should add that speech and writing are not only a means for handling and conceiving something in its symbolic aspect according to a general rule, but have all of the above-mentioned characteristics themselves:[2] They can be felt and experienced — or they can also be understood and used *as* signs associated with conventional meanings. And while conventional signs are shared among subjects, the experience of language is, among other things, the experience of its resistance to one's own particular associations between a sign and its meaning that are only partially reflected in common language usage, as well as the experience of verbalizing something in general. Thus situating oneself within language means varying and contextualizing it, as well as stabilizing and de-contextualizing an abundance of subjective ways of perceiving and dealing with something by referring to it with a sign. In doing so, we situate not only ourselves as linguistic subjects, but also language itself in so far as it is used in a specific context. Obviously, this is a two-way process in which both sides mutually constitute one another. Having a 'feel' for lan-

---

[2] 'There are three kinds of interest we may take in a thing. First, we may have a primary interest in it for itself. Second, we may have a secondary interest in it, on account of its reactions with other things. Third, we may have a mediatory interest in it, in so far as it conveys to a mind an idea about a thing. In so far as it does this, it is a *sign*, or representation' (Peirce 1998, p. 8).

guage, finally, means not just perception of its physical qualities but also the capacity to perceive something as language at all. Just as an alexithymic individual can see other peoples' facial expressions but is basically unable to associate them with any emotional state they might express, someone lacking the sense of language would be incapable of becoming aware of the symbolic dimension of a sign and would treat it instead as an object.[3]

Since we cannot leave the symbolic dimension behind, it would seem to make little sense to try to follow the historical genesis of our situatedness in language from sensation to experience up to its symbolic representation. The following sections will take the opposite path: starting from signs (2) and several general remarks on poetic speech (3) and following this back by way of the act (4) to the materiality of language (5), the chapter will seek to theoretically isolate these practically inseparable aspects and examine how they condition and relate to each other. Of course in examining one aspect we will not be able to entirely abstract from the other two.

2 Drawing distinctions

All three of these components — the sensible perception of language, the experience of the difference both between language and life-world and between one's own speech and the language of all, and its use as a means — are generally present simultaneously. However, in practice they are not equally present to awareness. In unproblematic communicative situations, the sensible qualities of language go largely unnoticed, and in most cases to focus on them would be contrary to the intentions of the participants. Someone who listens attentively to the intonation of a question rather than answering it will most likely be a disappointment to the questioner, and someone who attends a lecture on solar eclipses and spends most of their time enjoying hidden internal rhymes will do so at the cost of their understanding of astronomy. The material qualities of language are

---

[3] This happens, for example, to an illiterate shepherd in a dramatic fragment from Euripides, who finds an inscription on a rock in the woods. Not knowing that the oddly formed depressions are signs, he describes the letters by their geometric form and talks about them as if they were objects:
'a circle such as is measured out with compasses,
that has in its centre a conspicuous mark;
the second, first of all a pair of lines,
and another one holding these apart at their middles;
third, something like a curly lock of hair,
and the fourth has one part standing upright,
and three more that are fastened crosswise on it;
the fifth is not an easy one to explain —
there are two lines that begin from separate points,
and these run together into a single base;
and the last of all is similar to the third'
(Euripides, fr. 382,3, Nauck 1889, p. 477; engl. Collard and Cropp 2008, p. 421.)
He remains ignorant that it is the gravestone of Theseus he has come across.

only the subject of conscious attention when they get in the way. Asking someone else to please speak louder, slower, or more clearly (or using a magnifying glass to decipher a text) is only an intermediary step in order to then proceed with the business of communication. Also the experience of various ways of speaking is only relevant here when it threatens to impede communication and understanding—unless we change the subject and declare the linguistic difference itself the topic of conversation.

The debates as to whether these are in fact 'merely linguistic' differences look back on a long philosophical tradition. Plato, in his *Cratylus*, had already characterized words (or, more precisely: names) as 'an instrument of teaching and of distinguishing natures'[4] and thus considered language in its cognitive function and not just as an instrument of communication. Yet it is only with the growing insight into the historicity of meanings and the variety of languages—i.e. with the recognition that foreign languages are also *languages* and not just a barbaric stammering, as the ancient Greeks saw it—that people began to entertain the idea that languages are not merely modifications of an ideal structuring of our shared world, but rather belong to the various worlds that each language structures differently. What Wilhelm von Humboldt later called the *dividing activity of language* and André Martinet the *cutting up of experiential facts* means, among other things, that language both opens and obstructs our access to the world, at the same time—in disclosing it conceptually and yet always doing so in the particular manner of an individual language, which cannot so easily be matched up with other worlds through simple translational techniques. Moreover, both Humboldt and Martinet also saw quite clearly that this conceptual articulation can only be accomplished by its manifestation in and through language, which is bound to material signs that are themselves articulated. Unlike other systems of signs, language exhibits the so-called principle of double articulation: linguistic expressions can be divided into meaningful and distinctive units, i.e. they feature both morphemic and phonemic articulation. The linguistic articulation of thought is only possible due to the infinite possibilities of combination among a finite number of phonemes (Humboldt 1994; Martinet 1960).

The attribution of predicators makes it possible to deal with objects and affairs, but it also precludes other types of attribution. Given the plethora of subjective experience, the privileging of certain linguistic representations provides a valuable orientation but at the same time it severely confines the potential diversity of contextual relations—and this holds all the more, the more one distances oneself from the experiential contexts in question. Linguistic acts of distinguishing draw limits, and limits are painful. Singling out a schematic aspect from the inexhaustible wealth of

---

4 'Then a name is an instrument of teaching and of distinguishing natures, as the shuttle is of distinguishing the threads of the web' (Plato, *Cratylus* 388c).

situative experience can also be seen as an act of violence, and there could hardly be a more impressive expression of this than Rilke's poem about fear of the word:

> Ich fürchte mich so vor der Menschen Wort.
> Sie sprechen alles so deutlich aus:
> Und dieses heißt Hund, und jenes heißt Haus,
> und hier ist Beginn und das Ende ist dort.[5]

Yet in this poem Rilke does more than just bemoan the destruction of the life context and the violent limitation of the unlimited possibilities of perception — he also undermines them. In generating all sorts of new connections with sound and rhythm, he shows that language doesn't just divide but also joins — and that it is precisely the *movement* of dividing and connecting that offers us the possibility of redemption from rigidified, formulaic articulations. The poet does not merely suffer from bumping into the borders of language, but vividly illustrates it, conscious of his own situatedness. However, turning against this sufferance does not mean that in doing so he imagines himself to have entered entirely into the realm of action. Rilke does not claim, in the form of an assertion, to suffer language and yet to be able to free himself from it through his own speech. Rather, he presents this in the medium of intuition, and thereby allows us as well to undergo the experience of an aesthetic difference, for example in our perceiving that hounds, in terms of sound, have much more in common with houses than with other four-legged creatures, or in our noticing the discrepancy between how the language of the representation weaves and glides compared to the stiffness of the language represented. Whether we simply yield to these experiences in immersing ourselves in Rilke's world of sound, or whether we begin to examine the schemata of sound and rhyme, might depend on the interest and analytic capacities of the particular reader. The crucial point is that without truly experiencing the poem for oneself physically (for example if one merely registers that an alliteration is a stylistic device whereby the root syllables that are emphasized in two or more words have the same initial sound, and that by means of this stylistic device we can generate the most astonishing connections) we could not relate this to our experiences in any way — in which case we could in fact speak of non-situated aesthetics.

A passage from Daniel Kehlmann's novel about Alexander von Humboldt's research expeditions illustrates, with charming irony, just how distinct the creation of a poetic object is from the mere registering of a state of affairs by conventional means. When Humboldt is asked, during a

---

5   'The words of humans fill me with fear.
    They name everything with articulate sound:
    So this is called house and that is called hound,
    And the end's over there and the start's over here'
    (Rilke 1996, p. 106; engl. translation by Walter A. Aue.)

boat trip on the Rio Negro, to tell a story, he declines, offering instead to recite 'the most beautiful German poem', Goethe's *Wanderers Nachtlied*:

> Here it was. Above all the mountaintops it was silent, there was no wind in the trees, even the birds were quiet, and soon death would come. Everyone looked at him. That's it, said Humboldt. Yes, but, asked Bonpland. Humboldt reached for the sextant. Pardon, said Julio, but that couldn't have been the whole thing. (Kehlmann 2007, p. 107)

What the narrator, more interested in measuring and weighing than telling stories, has withheld from his baffled listeners is more than just the sensory dimension of *Wanderers Nachtlied*. The episode concerns native raconteurs of magical stories far removed from all European traditions of thought, who had hoped to learn something about foreign ways of telling stories from the German Humboldt. This would also have presented an opportunity to learn something about Humboldt himself from his way of narrating. Yet he not only refuses to tell a story, promising a lyrical presentation instead, but then, the consummate scientist, reproduces the contents in matter-of-fact Spanish. One could hardly get any farther away from poetic speech. The audience, deprived of both the phenomenal experience of the poem and the possibility of experiencing part of a foreign world, justifiedly feels cheated: 'pardon?'[6]

3 Dancing in chains

Situating oneself linguistically means more than just becoming aware of the sensory aspect of speech or writing—beyond all spatial metaphors about the 'inner' and the 'outer', it also involves actively placing oneself in a relation to the framing conditions that disclose the spaces of possibilities for one's linguistic action and their limits. These sorts of framing conditions could be language in general as well as the rules and conventions of a particular language and even sociolects or idiolects. Whereas we usually experience them involuntarily in everyday communication (as can manifest itself in searching for the right words, unintended grammatical errors, or in the sort of elimination of obstacles to communication mentioned above), in literature and poetry we are concerned with the conscious exploration of linguistic liminal experiences, which can sometimes be intensified by additional self-imposed rules. Nietzsche calls this form of

---

6   For those interested in the sensory side of the poem, the original version runs as follows:
    'Über allen Gipfeln
    Ist Ruh'
    In allen Wipfeln
    Spürest du
    Kaum einen Hauch;
    Die Vögelein schweigen im Walde.
    Warte nur, balde
    Ruhest Du auch.'
    (Goethe 1982, p. 142.)

self-limitation and the playful overcoming of it a 'dance in chains' that brings out all the more clearly poetic freedom as the freedom of a prisoner as well as the nature of the imprisonment.[7] We need only remember the subjugation to complicated metres and rhyme schemes, or George Perec's *Disparition*, a novel written without the most common grapheme in the French language, the letter 'e'.

The effect of this sort of approach is to shift attention from the semantic contents conveyed to the linguistic means and the form of organization of the text. Roman Jakobson defined the poetic function of language as the transfer of paradigmatic criteria of selection to the syntagmatic axis.[8] This means that linguistic units that stand in some relation of similarity or dissimilarity to one another, and thus represent mutually exclusive alternatives, end up in a relation of proximity—which both suspends the ordinary principles of organization and accentuates them. Thus the use of the word 'child' in a certain position in the sentence precludes the alternatives 'minor' and 'old man', in looking for a suitable preposition one chooses either 'over', 'between' or 'under', and in the paradigm of words beginning with 'h' one chooses between 'hound' and 'house', depending on context. In poetic speech, however, it is precisely the similarities and differences that determine the linear arrangement: synonymical and antonymical sequences, end-rhymes and the breaking out of the rhyme scheme, equal and unequal syllable counts, assonances and dissonances, and regularities and irregularities of emphasis become constitutive for the syntagmatic level and thus make the paradigmatic level transparent. The effect of juxtaposing linguistic units that normally preclude and conceal each other this way is that the points of intersection in the dividing activity of language suddenly appear fresh—and revisable.

4 Difference in motion

The more the 'focus on the message for its own sake' (Jakobson 1960, p. 156) in such procedures pushes the sensible dimension of language into the foreground, the more they challenge us to situate our own physical experience in the production and reception of language as well.

The rhythmic composition of a text is a formative element that is connected in a special way to physical experience. Characterized by Ludwig Klages as the reproduction of the similar within sameness, it is distinct from both regular units of metre and the pure sound material, just as the

---

[7] '"Dancing in chains", making things difficult for oneself and then spreading over it the illusion of ease and facility... firstly to allow a multiplicity of constraints to be imposed upon one; then to devise an additional new constraint, impose it on oneself and conquer it with charm and grace: so that both the constraint and its conquest are noticed and admired' (The Wanderer and His Shadow 140, in Nietzsche 1996, p. 343).

[8] 'The poetic function projects the principle of equivalence from the axis of selection into the axis of combination' (Jakobson 1960, p. 358).

individual use of a linguistic expression differs from the general meaning and the sound that expresses the subject's sensation. In regard to the fluid (and captivating) affective movement of sound, the rhythmic articulation represents a moment of inner freedom; within the evenness of metre this subjective integration arises as individual character.[9] However, whereas the semiotic structure divides the world into conceptual units, rhythm divides time and thus moves poetic speech closer to music and dance: rhythmic articulations make time experiencable in human measure — taking in both mental constructs and physical sensations.[10]

The French linguist Émile Benveniste, in his work of conceptual history 'La Notion de "Rhythme" dans son Expression Linguistique', divorced rhythm from its typical derivation from the Greek 'rhéin' (flow) as well as from the concept of 'schéma' as objectified form, and due to its ending (th-)mós, which indicates 'not the completion of a concept, but the particular modality of its completion in intuition' (Benveniste 1966, p. 332) characterized it as 'the form in the moment when it is assumed by that which is mobile, moving, fluid' (*ibid.*, p. 333). The ancient musical theorist Aristoxenus of Tarentum, whose work only survived in fragmentary form (Pearson 1990), had already made perceptibility the measure of the smallest rhythmic unit.[11] But at the same time he defined this unit as a duality: to speak of a rhythmic unit requires at least one heavy and one light syllable — so it is more a unit of an act of distinguishing. The result of such distinctions is that, on the one hand, we can now deal with the objects they have won for us (for example, we can speak of heavy and light syllables and perceive or pronounce them), or we can also 'make the distinction as such the object of a further distinction' and thus make it 'one side of another distinction' (Hagen 2009, p. 83).[12] Yet we can also focus on the *act* of distinction rather than its results and experience it as a certain form of movement within a scheme of metre or in contrast to other forms of movement.

We are only able to perceive rhythm at all, as the qualitative variation of either a regular change of emphasis or purely quantitatively determined temporal structures, due to our ability to generate regularities in memory and anticipation. In this context Ernst Gombrich speaks of a *sense of order* (Gombrich 1984, p. 289), comparable to the ability to conceptualize our experiences in light of a self-discovered general rule that Peirce called the

---

9 *Cf.* Dahlhaus (2002, IV, p. 558).
10 'Only in poetry with its regular reiteration of equivalent units is the time of the speech flow experienced as it is — to cite another semiotic pattern — with musical time' (Jakobson 1960, p. 358).
11 Also Klages locates rhythm at the phenomenological level: 'Rhythm belongs to the world of appearances' (Klages 1934, p. 7).
12 This would be the case, for instance, when a consistently maintained trochaic verse is given a lighter note by the insertion of dactyls, or countered by dramatic semantic development.

sense of learning (Peirce 1998, p. 8). And just as the *sense of learning* saves us from being subjected to a disjointed series of competing sensations, the *sense of order* equips us to subsume sequential moments into larger unities and to conceive their subjective characteristics as the return of the same in different form. Thus at the level of the articulation of time, rhythm takes on the systematic role that individual speech occupies within the general semiotic articulation of experience—and like the latter, it can emerge as unintentional or conscious variation, as a situating that we merely undergo or that we reflect on (in advance or retroactively). In contrast to the general structures in language as a system, rhythmic temporal sequences and individual speech acts exhibit a kind of perpetual articulation, in the formation of a distinction in the indistinct as well as in the already distinct, which allows the *processes* of formation and perception to be foregrounded rather than their results.

Yet the framing conditions of rhythmic structure in general include other rhythms besides the specifications of measure or metre. One's own work rhythm relative to the shift from day to night, the structuring of a day of vacation on the ocean in view of the tides, the rhythm of two people finding and then losing each other against the background of the change of seasons, the rhythm of a poem relative to one's heartbeat (and also the heartbeat at various times of day)—all of these are overlapping qualitative forms of experiencing time that can reinforce or lessen, support or disrupt each other. Alongside these natural periodizations, we are also confronted with the idiosyncratic rhythms of other people. These can have a natural character as well (for example if someone speaks so sluggishly and monotonously from weariness that we can no longer follow him and lose ourselves in our thoughts) or they can also be intentional (for example when someone skilfully manages to interrupt us again and again at decisive points so as to maintain the upper hand in the conversation, or when an actor gives a verse of Shakespeare additional allure with her own temporal articulation). An aesthetic experience only takes place—or, as Peirce would say: the surprise is only perfect—when the percipients do not ignore their own rhythm(s) while perceiving and considering rhythmically structured processes and acts. Situatedness begins on the variable border where the question of identity and alterity is always posed anew.[13]

5 Una voce poco fa[14]

Several authors, Klages above all, have entirely removed rhythm from the field of symbolic acts and assigned it instead to the realm of nature—a

---

13 On learning from one another as a process of continuous distinctions between one's own and others see Lorenz (1998).
14 Una voce poco fa/qui nel cuor mi risuonò [A voice has just echoed/here into my heart] (Rossini, *Il barbiere di Siviglia*).

nature that, as momentary form, stands out as a symptom and does so all the more clearly the more it is held in certain limits by the mind (the measuring and counting mind of natural science). Without entirely sharing this view, we cannot deny that the temporally bound character of language is normally only an accompaniment, displayed by language rather than depicted within it. Yet the poetic turning back of language upon itself as living and subject to time, as the 'mirror of human life among other things' (Klages 1934, p. 46) illustrates this temporally bound nature and thus allows a reflection upon one's own transience.

We see this not least of all in the fact that the phonetic material is continually passing as it emerges. Resounding and fading at the same time, the human voice is irretrievably lost in its presence.[15] Yet it is precisely as voice that it refers us not just to the living (and mortal) body that produces it, but also to the *possibility* of speech — its materiality is the materiality of a sign. Roland Barthes called this erotically charged contact surface, where the corporeality of the voice and language rub against each other, the 'grain of voice'. Beyond communicative or referential aspects and subjective expression, the 'volume of the singing and speaking voice… is the space in which meanings germinate "from within language and in its very materiality"' (Barthes 1977, p. 182). Where sound and meaning emerge together, the phonemic level is no longer subordinate to the morphemic one; rather, it is the rasping of consonants, the making audible of the processes of articulation on the tongue, teeth, and gums and the resounding of the vowels from out of the depth of the body that makes the voice seem desirable; and yet this pleasure is a pleasure in the sign, that hopes for and presupposes a sense of language.

The experience of the materiality of language is endangered and limited from two sides: either when one tries to fix it semiotically like a pinned butterfly (a futile endeavour, as Barthes has shown using the example of efforts to recall a voice that has died away through unending series of predicates — 'agile, fragile, youthful, somewhat broken?' — Barthes 1994, p. 67). Or one naturalizes it, and gives oneself over entirely to the inexpressible sensation of a meaningless sound that does not promise speech and ultimately ends up silencing both parties. As it concerns the situating of language, this corresponds firstly to an exorbitant reflectiveness that tries vainly to evoke (and to avert) the physical presence of the other as well as one's own physical experience, and secondly to merely suffered (or enjoyed) abandoning of oneself to an unarticulated and unarticulatable 'something' beyond all distinctions.

As an aesthetic programme the exploration of the sensual aspect of language is most pronounced in concrete poetry, which suspends the referen-

---

15 Roland Barthes writes: 'The voice is always *already* dead, and it is by a kind of desperate denial that we call it: living' (Barthes 1994, p. 68).

tial function of language through its focus on its acoustic and visual aspects, and in *poésie pure*, which claims neither to express subjective feelings nor to demonstrate an extra-linguistic reality. Yet the exhibition of this tension between nameless sensory experience and the ceaseless attempt at predication ultimately takes its poetic foundation from Romantic considerations of poetry as a longingly circling around the unsayable, in a language able to reflect the play of things in a play with itself (Novalis).

Such an artistically reflected situatedness can uncover the physical, phonemic aspects of language just as well as it can reveal the fact that objects are never directly accessible to us, but only through symbolic mediation. We cannot get them completely under our disposal through conceptual structures and distance ourselves from all experiential contexts: this would neglect the fact that we not only confront the world as speaking and acting subjects but are also a part of it. Neither can we leave the symbolic dimension entirely behind us and see ourselves as merging with the world: without a *sense of reaction* and *sense of learning* we would disappear into it and it would disappear with us.

We can only experience this dismantling of semiotic orientations by turning back upon the material properties of language in light of our situatedness as we reflect upon it. Becoming aware of the uniqueness and transience of one's own and other's speech is only possible for someone who knows the difference between you and I, here and there, then and now, life and death. And yet fails to grasp it again and again.

References

Barthes, R. (1977), *Image – Music – Text*, essays selected and trans. S. Heath, London, Hill and Wang.

Barthes, R. (1994), *Roland Barthes by Roland Barthes*, R. Howard, Trans., Berkeley (CA), University of California Press.

Benveniste, É. (1966), 'La Notion de "Rhythme" dans son Expression Linguistique' in *Problèmes de linguistique générale*, vol. 1, Paris, Gallimard: 327-35.

Dahlhaus, C. (2002), *Gesammelte Schriften in 10 Bänden*, Bd. IV, *19. Jahrhundert I: Theorie/Ästhetik/Geschichte, Monographien*, H. Danuser, Ed., Laaber, Laaber Laabererlag.

Euripides, fr. 382, in A. Nauck, Ed., *Tragicorum Graecorum Fragmenta*, Bd. 2, Leipzig 1889. Engl. (2008), C. Collard and M. Cropp, Eds. and Trans., *Fragments*, Boston (MA), Harvard University Press.

Goethe, J.W. von (1982), *Werke*, Hamburger Ausgabe in 14 Bänden, Bd. 1, *Gedichte und Epen I*, Hamburg, Hanser.

Gombrich, E. (1984), *The Sense of Order: A Study in the Psychology of Decorative Art*, London and New York, Phaidon Press.

Hagen, W., Ed. (2009), 'Gibt es Kunst außerhalb der Kunst? Niklas Luhmann im Gespräch mit Hans-Dieter Huber' in *Was tun, Herr Luhmann? Vorletzte Gespräche mit Niklas Luhmann*, Berlin, Kulterverlag Kadmos.

Humboldt, W. von (1994), 'Über die Buchstabenschrift' in J. Trabant, Ed., *Über die Sprache: Reden vor der Akademie*, Tübingen/Basel, Utb Gmbh: 98-125.

Jakobson, R. (1960), 'Closing Statement: Linguistics and Poetics' in T. Sebeok, Ed., *Style in Language*, Cambridge (MA), MIT Press: 350-77.

Kehlmann, D. (2005), *Die Vermessung der Welt*, Hamburg. Engl. (2007) C.B. Janeway, Trans., *Measuring the World*, New York, Quercus Publishing.
Klages, L. (1934), *Vom Wesen des Rhythmus*, Kampen auf Sylt, Kampmann.
Lorenz, K. (1998), 'Einleitung: Das Fremde und das Eigene im Vergleich' in *Indische Denker*, München, Beck: 15–32.
Martinet, A. (1960), *Eléments de Linguistique Générale*, Paris, Armand Colin.
Nietzsche, F. (1981), 'Sämtliche Werke' Kritische Studienausgabe in 15 Bänden, G. Colli und M. Montinari, Eds., KSA 2: *Menschliches, Allzu Menschliches* II, Der Wanderer und sein Schatten. Engl. (1996), R.J. Hollingdale, Trans., *Human, All Too Human II, The Wanderer and His Shadow*, Cambridge Texts in the History of Philosophy, Cambridge, Cambridge University Press.
Pearson, L. (1990), *Aristoxenus: Elementa Rhythmica. The Fragment of Book II and the Additional Evidence for Aristoxenean Rhythmic Theory*, Greek texts with intro., trans. and commentary, Oxford, Clarendon Press.
Peirce, C.S. (1998), 'What is a Sign?' in Peirce Editions Project, Ed., *The Essential Peirce: Selected Philosophical Writings*, Vol. 2 (1893–1913), Bloomington (IN), Indiana University Press.
Plato (1990), *Werke*, 8 Bde., Griechisch und Deutsch, G. Eigler, Ed., Bd. 3: Phaidon. Das Gastmahl. Kratylos, Darmstadt.
Rilke, R.M. (1996), *Gedichte 1895 bis 1910*, Kommentierte Ausgabe in 4 Bänden, Bd. 1. Engel, M., U. Fülleborn, H. Nalewski *et al.*, eds, Frankfurt am Main/Leipzig, Insel.
Schneider, H.J. (1992), 'Zur Einführung: Der Begriff der Erfahrung und die Wissenschaften vom Menschen' in H.J. Schneider and R. Inhetveen, Eds., *Enteignen uns die Wissenschaften? Zum Verhältnis zwischen Erfahrung und Empirie*, München, Fink: 193–210.

Paola Carbone

An Externalist Approach to Literature
From Novel to Cave Writing

My purpose is to investigate to what extent the externalist stance might prove useful in exploring both the way a writer enacts a fictional world, and how the reader makes that fictional world real for herself. Instead of talking in the abstract, I intend to proceed through a reading of several well-known English novels and digital narrations from the eighteenth century through to cave literature. Such a huge overview might seem a challenging endeavour, yet irrespective of contrasting aesthetic perspectives, literature has always been aware of the unavoidable and vital relation between subject and object or between living beings and environment. I will champion the view that the dichotomy between objective and subjective reality is nothing more than a disguised and misleading version of the underlying union of subject and object expressed by the interplay between language and narration. How has literature developed these relationships? What is the reader's role?

1 *Tristram Shandy* by Laurence Sterne: How experience becomes a world

When the novel made its major breakthrough in the eighteenth century — during the age of Enlightenment, science and measurement — the purpose of the novelist was to create a *sense of reality* so as to facilitate the reader's identification of the fictitious world with the real world in which she lived. Therefore, cognitive bridges were erected between fiction and reality in order to create low-mimetic representations: characters were sketched out assigning names and surnames, setting them in a specific and well-defined temporal and spatial dimension and developing a series of adventures based on cause-and-effect relationships. The main character was supposed to be an *exemplum* for the 'new' bourgeois citizen, who was engaged in experimenting with her role in society. Although the character is conceived as a single individual, she is presented as in a continuum with the environment: she does not exist without the social context and *vice versa*. Through the main character, *experience becomes world*, because her entire existence, values, and morality depend on the physical, intellectual and emotional (I mean, experiential) interaction with the context.

As a first example, nothing is better than *The Life and Opinions of Tristram Shandy, Gentleman* by Laurence Sterne: a chaotic, digressive, non-linear masterpiece. It is a borderline novel steering between realism and avant-garde. While it addresses all the cultural facets and the main debates of the age, it challenges both the eighteenth century idea of an objective knowledge and the belief in the separation between an objective reality and the subjective identity. What makes this novel unique amongst its peers is the way in which the *process of writing* coincides with the *making* of Tristram and his own world. Here, instead of an *imitatio naturae* as in realist literary works, the reader is present at the very moment that a subject and his world are *generated*. The conventional effort that characterizes the typical realist *mimesis* (inspired by the Aristotelian *adaequatio rei et intellectus* oriented towards an objective and autonomous reality) resembles the use of perspective in painting. Sterne does something totally different and unexpected linking inseparably subject and object, reader and character, experience and reality.

The narration follows the point of view of the narrator—Tristram, heir of Shandy Hall—who develops his consciousness and sensitivity while he forges his *story*. As Locke theorizes, the young protagonist starts out as a *tabula rasa* and then constructs his *being* through physical exchanges with the environment, memories of past experiences, and rational thoughts. The title tells us that the novel is about his 'life' and 'opinions': the continuous connection of ideas either intuitive or habitual (Locke 1690, Book IV) underpins his entire phenomenological activity to such an extent that the narration is a collection of apparently incoherent mental associations. Step by step, perceptual experience and rational knowledge guide him towards the development of his opinions, even though his self-awareness as *un soi-même* never emerges explicitly.

An externalist perspective appears to be more than just latent here and for more than one reason. First of all the *peripetia* is enacted by colliding personalities. These disagreeing identities fail to share a common first-person perspective not because they are peculiarly different, but rather because they have a divergent knowledge of what is 'real' since they experience the same world in different ways—they own different *Umwelt*. If being conscious entails perceiving, thinking about, and acting in the world, Sterne gives to each single character's consciousness the hallmark of a mental-dependency called *hobby-horse*, i.e. an interpretative key for reality. For instance, Uncle Toby 'depends' on (he sees, reads, interprets whatever happens to him in terms of) war and artillery. Walter Shandy depends on philosophy, names, and noses. Mrs Wadman on Uncle Toby and sex, and so on. Paradoxically, *hobby-horses* function as explicative gaps. Since each character perceives and generates reality in a peculiar way, hobby-horses become a space of commonly acknowledged misunderstanding. Consider when Walter Shandy gives one of his long speeches and his brother Toby mixes up the semantic fields of his words so that a *train* of ideas becomes a *train* of artillery. Since Walter knows about Toby's obsessive metaphors, although he gets angry their communication is, in any case, successful. *Tristram Shandy* is an array of living microcosms in continuous interaction. Each character is neither contained inside a private mental world, nor isolated from either the literary or actual world. Each character singles out a part of the world.

The author transforms these collisions between worlds into a source of amusement, training the reader to grasp Sterne's irony and subtle innuendos. From the very first page, these forms of (mistaken-in) communication enact the world of Shandy Hall in front of the reader whose awkward task is to keep all the separate entities together inside an integrated epistemological context. The reader is not just the detached observer of the adventures of a group of fellows. On the contrary, he is urged on to discover cognitive unities in the apparent disorder of words and ideas. He is asked to recognize a complex of meanings based on multiple interactions between animate and inanimate *dramatis personae* (Carbone 2008), so that the more complex the net of relationships, the less the work appears to be disordered and chaotic.

Just as Tristram experiences his life, Laurence Sterne asks the reader to become a multi-task *performer*. As a matter of fact, the author exploits a complex verbal-visual apparatus made of black pages, marbled pages, white pages, lines, dashes, asterisks, etc. in order to break down the stability of the book's printed page. He prefers a dynamic multimodal reading typology to a linear reading praxis. His purpose is to reveal the *trompe-l'œil* hidden in the realistic narration—namely that sense of reality which is nothing but the construction of a single ordered and intelligible world. To avoid such a deceitful perspective, Sterne demands a collaborative attitude and a total commitment to the text, as the narrator often reminds and warns the reader.

The fragments of *Tristram Shandy* are like the colours of a painting perceived by the eyes of an observer. The colours are not just pigments on a canvas. The scenes are not just ideas emerging from the characters' minds, but rather sensorial extensions of the reader in relation with the book, the author, and the characters. The reader singles out only those connections that she can recognize. The text acts on the reader and *vice versa*. There is a unity. The reading performance becomes a physical, perceptive, and cognitive act, inseparable from the reader's cognitive universe. It is not just a matter of receiving information about life at Shandy Hall. The reader's consciousness brings the entire Shandean community actually into existence (Honderich 1998, p. 140).

Even if Sterne does not deny the prevalent eighteenth century philosophical assumptions, an externalist perspective seems to be deeply rooted in his work. Taking advantage of a subjective and ambivalent reality, the novelist and the characters provide the reader with multiple events, opinions, and meanings she is supposed to re-organize into a unity. Such a unity is often and in several different ways solicited to offer an emotional, intellectual and cognitive experience as to the transfiguration of Shandy Hall into an actual world.

2 *Daniel Deronda* by George Eliot: Physical experience as knowledge

The internalist model of the mind conceives our experience as nothing more than the brain's interpretation of incoming information. However, so far, no empirical confirmation of this model, which does not explain the *fons et origo* of the meaning associated with the information, has been provided. Generally speaking, the physical world is necessary for initiating a neural event, but physical perception alone is certainly not enough for telling a story. If it were, the narration itself would be nothing more than the description of a series of stimulations and physical responses. Of course, a story needs a conscious reaction to situations to guarantee a narrative development, since a neural process as such cannot give meaning to any kind. Such an assumption would be tantamount to falling into the mereological fallacy (Bennett and Hacker 2003). In this sense, literature proves useful in demonstrating how the externalist view of the mind is able to locate meaning in a broader network of relations and interactions.

Nineteenth century novels give us the opportunity to investigate exactly *how* a character becomes conscious of the world, seen not as an absolute entity-in-itself (Honderich 1998, pp. 148–55), but as *a perceived reality dependent on her physical experience*. During the nineteenth century, the character was no longer an *exemplum*, but rather an individual reaching mental and emotional maturity by means of life challenges. Similarly, fiction conceives subjectivity as a matter of developing and growing a personality. Up to the end of the eighteenth century, self-control and rationality were hailed as key virtues, since sense, together with strength of mind, meant 'right judgment', 'common sense' and 'rationality' — all qualities that were essential to preserve one's own virtue (especially in women's case) and reputation. Characters struggled to hide their emotions because they felt that emotional transparency meant vulgarity and light-mindedness. When, under the influence of J.-J. Rousseau, the balance between personality and decorum turned into the cult of sensibility and naturalness, sentimentalism became the perfect tool against rationalism and modern society. Subsequently, social intercourse measured the intensity of emotion and moral value in visible demonstrations of feelings, whereas, in the arts, sensibility was used to stir up the passions of spectators and readers. The work of art became a psychological expression whose value lay in the immediacy and uniqueness of the individual experience described.

As Henry James pointed out in 1873 (Raphael 2001, p. 59), George Eliot was the first English writer to take the English novel a step forward with the introduction of psychological realism. For her, the setting of the story lies mainly in the minds of the characters and deals with the analysis and interpretation of reality.[1] The novel takes form from the physical and instinctual relationship that the character establishes with the world (and *vice versa*). Eliot's realism is based on the assumption that emotions can take wide-ranging forms since experience is made up of perceptions embodied in the 'fibres' under the skin. It is through the skin that character and context are brought to the reader's knowledge: the hues of society are revealed by different minds through perceptive processes made up of bodily (re)actions. Eliot's purpose was not the representation of things as they should be, but the description of subjectivity through the ever-present physicality of mental processes (Davis 2006, p. 28).

In *Daniel Deronda* (1876)[2], the author presents facts which begin as perceptive moments derived from neural, chemical or physiological reactions of the body to the environment. This must not be critically confused with a straightforward internalist position. On the contrary, I venture to point to an externalist logic implicit in the apparently internalist context. Unlike Tristram's digressive and *in-process* self, Eliot's characters are adults who develop their identity under the pressure of the environment. The opening of the novel is emblematic:

> Round two long tables were gathered two serried crowds of human beings, all save one having their faces and attention bent on the tables… Deronda… was… arrested by a young lady who… showed the full height of a graceful figure, with a face which might possibly be looked at without admiration, but could hardly be passed with indif-

---

[1] The heated debate on natural selection pushed many thinkers and scientists to discuss about this relation. In 1868, Thomas Henry Huxley explained how physical processes cause mental processes, rather than the other way round, affirming that 'thoughts are the expression of molecular changes in that matter of life which is the source of our other vital phenomena' (Huxley 1869, pp. 130–65). According to him, a subjective mind can be described only in physical terms and through the observation of the body. Concurrently, George Henry Lewes — philosopher, scientist and husband to Marian Evans — claimed a specific language was necessary to speak of subjective psychological experience, as Freud would eventually provide.

[2] The novel contains two subplots. The first subplot deals with a stubborn, selfish and beautiful woman called Gwendolen who makes her own way in the world with satirical observations and manipulative behaviour. The second one focuses on Daniel Deronda, a light-hearted and compassionate young man who rescues from suicide and marries Mirah, a young Jewess.

ference... But in the course of that survey her eyes met Deronda's... The darting sense that he was measuring her and looking down on her as an inferior, that he was of different quality from the human dross around her, that he felt himself in a *region outside and above her*, and was examining her as a specimen of a lower order, roused a tingling resentment which stretched the moment with conflict. It did not bring the blood to her cheeks, but it sent it away from her lips. *She controlled herself by the help of an inward defiance, and without other sign of emotion than this lip-paleness* turned to her play... [S]he felt the orbits of her eyes getting hot, and the certainty she had (without looking) of that man still watching her was something like a pressure which begins to be torturing. (Eliot 1876/1970, pp. 35–9, my italics)

In the gaming room the two characters appear as the only beating living beings. Here, sight endorses the perceptual content as the most important among the senses. Experience is eventually enacted through thoughts in each individual (Noë 2004). The narrator underlines the characters' capacity to perceive and react to the environment so as to enact their own living experience. Their catching each other's eye is the perceptive response to the physical presence of both of them in the same place. From the very first glance, they influence each other so that Gwendolen feels observed by a man whom she suspects feels himself superior to her (as if he were 'outside and above her') and this affects what she believes and the way she remains in the room. They both give and receive sensory stimulus information which reflects off their skin, turning their bodies into texts open to personal interpretation. While she reacts physically to the stimuli around her, he looks at her as if observing a scientific experiment. A sensorimotor process takes place involving the entire body. It is supported by the rational thought which, in turn, leads to yet another physical reaction: the woman's 'inward defiance'. In a way, the two characters are complementary to one another as a result of both environment and conscious behaviour. They are totally coupled together by means of the continuity offered by the environment.

From the above passage, we may infer that Gwendolen's receptors have been stimulated even if we do not know exactly what was happening in the room. Through her point of view and her reactions, we can imagine the whole scene, filling in the gaps as usually happens by means of *amodal perception*. Recurrently in Eliot's novels there is the feeling of possessing a complete awareness of the story while, in reality, the effects are sketched without any explicit and objective reference to the cause. The reader feels involved in the story told because of the sympathy and empathy of the perceptive activity presented. Furthermore, the emotions and impressions stimulated are part of the novel because the reader needs emotional involvement to feel the plot as real. It is interesting to note that the reader sometimes *sees* differently from the character (that is she has a superior knowledge of the facts), but this cognitive capacity is foreseen by the author and integrated into the meaning of the story-telling. Such a twofold awareness allows the reader to understand the development of both character and narration as they are experienced through the act of reading.

In short, it would apparently seem that an external stimulus affects Gwendolen's state of mind and that her own flesh reacts instinctively, revealing her internal states to the outer world. In any case, *inner* and *outer* are virtual categories since each character decides how to behave according to her particular experience and consciousness. Inner and outer are two essential but relative spaces constantly in communication.

James J. Gibson (1979, p. 135) has pointed out that a human being reflects information on the skin in order to communicate what she is, invites, promises, threatens, or does. According to him, perception implies the presence of a meaningful and valuable object because all perceptions are underpinned by affordances which either benefit or injure others. So, the glance Deronda casts at Gwendolen is charged with moral judgment and intellectual value (at least that is what the woman understands) and it amplifies the opportunity for the story to develop. The glance is an emotional and physical affordance enacted by Gwendolen. From a narrative point of view, the novel suggests that thoughts are less important than the characters' reactions to the world, because it is only through feedback that thoughts can, on the one hand, become interesting for the reader and, on the other, give form to the content. Thoughts are not interesting in isolation.

While in Sterne we can see the making of a contextualized and situated consciousness, in Eliot's work we are confronted with the development of perceptions. In the former, the reader weaves a net of relationships among the objects of the story; in the latter, it is the physical appearance that becomes a source of knowledge which singles out the character in a social and cultural *milieu*. In both cases, it is the reader who brings these relations into being and, equally, those relationships constitute the reader.

3 Virginia Woolf and the luminous halo

At the beginning of the twentieth century, modernist literary poetics implicitly embodied the externalist stance as to the relationship between subject and object. Unlike most naturalist authors, modernist writers are interested in *vivid human beings* — in *free human individuals* rather than *social beings*.[3] The immediacy of representation can thus be conveyed only by consciousness and renewed literary conventions such as internal monologue, stream of consciousness, free in/direct discourse, fragmented verbal articulation, discontinuity, collage, multiple points-of-view, disruption of

---

3 These are expressions used by D.H. Lawrence referring to John Galsworthy's characters.

logical and chronological linearity.[4] Moreover, instead of one single reality, writers believe in the existence of as many realities as individual consciousnesses. As a consequence, their attention focuses more on the heterogeneous moments of unities between human beings and their environment rather than *what* characters do — objective facts are replaced by the experience of single individuals. Nothing exists outside the experience which is both world and consciousness. The form of the novel consists in the overlapping of such several consciousnesses. According to Henry James:

> Experience is never limited and it is never complete; it is an immense sensibility, a kind of huge spider-web of the finest silken threads suspended in the chamber of consciousness, catching every air-borne particle in its tissue. (James 1884/1948)

James recognized that, instead of the *world*, the *experience* is suspended in the chamber of consciousness, leading us to believe that fleshing out as a process a direct and constitutive relationship between subject and object is one key to realism. Likewise, according to Virginia Woolf, from the very beginning of life, consciousness is solicited by stimulating impressions the author is asked to represent:

> Examine for a moment an ordinary mind on an ordinary day. The mind receives a myriad impressions — trivial, fantastic, evanescent, or engraved with the sharpness of steel. From all sides they come, an *incessant shower of innumerable atoms;*... so that, if a writer were a free man and not a slave,... there would be no plot, no comedy, no tragedy, no love interest or catastrophe in the accepted style, and perhaps not a single button sewn on as the Bond Street tailors would have it. Life... is a *luminous halo, a semi-transparent envelope surrounding us from the beginning of consciousness to the end*. Is it not the task of the novelist to convey this varying, this unknown and uncircumscribed spirit, whatever aberration or complexity it may display, with as little mixture of the alien and external as possible? We are not pleading merely for courage and sincerity; we are suggesting that the proper stuff of fiction is a little other than custom would have us believe it. (Woolf 1925/2002, pp. 146–54, my italics)

The writer is engaged in the reordering of the introspective data that the subject processes in a continuum with the world. The transitory reality and the self are indistinguishable. Time and space are no longer objective. They are no longer measurable cognitive structures conventionally used to understand reality. They are cognitive categories *connecting the character with* her own world.

Influenced by Henri Bergson, Virginia Woolf writes *Mrs Dalloway*[5] (previously called *The Hours*) in which the two protagonists live through an

---

[4] As Pericles Lewis (2007) points out, modernism is not only a rebellion against realism or the Victorian novel, but it is the 'natural continuation of a trend' started by Dostoevsky, Flaubert and James.

[5] The novel is about two mirror-characters — Clarissa Dalloway and Septimus Warren Smith — who never meet, but one day in London share similar thoughts about life and events. In the end

ordinary day while Big Ben strikes the hours. The novel begins with 50-year-old Clarissa who perceives the fresh air in the morning in London only to find herself, in her mind, back at the family's summerhouse at Bourton, when she was young and desirous of love. While Woolf introduces the woman suggesting images of her present actions, she also lets the character experience her-*self* in relation to past and present experiences. Past experiences are vividly called up by present external reality to highlight the woman's being. However, the past is not a source of knowledge as it was for Tristram Shandy, nor is it forever over, just as Clarissa is not *simply* remembering her past youth. On the contrary, we can say that her past is still having an effect on her (Manzotti 2006). Clarissa's environment is extended spatially and temporally to all those events that are causally responsible for her bodily states. The fresh air in today's London is no more causally efficacious than the family's summerhouse at Bourton. The fresh air comes to the fore for the character and the reader not only because it is a physical occurrence in present time, but because it is the effect of a reminiscence. Otherwise it would have never been perceived or emphasized by the author. Thus memory is the literary and physical cause of the presence of a window at Bourton in the past.

The author no longer represents how things are, but *how things are in so far as they are experienced*. Experience and reality gets closer and closer without necessarily reducing one to another. No longer based on the assumption of the *mimetic equivalence* of fiction and reality, a new sense of reality emerges as a transitory unity. But the existence of such 'unity' depends on a particular subject who is able to recognize its existence. As a matter of fact, Clarissa's interior monologues take form from an interaction of external reality with the self: inner and outer reality merge into a single dimension. For the reader there are no barriers between the character and her surrounding, as the surrounding is an effect of the character's perception and awareness of it.

In Woolf's same work, similar examples are offered either by the black car with the drawn curtains or by the aeroplane which writes letters of smoke in the sky. Both are experienced in different ways by different people walking in the street. This shows that no single reality exists, and also that reality does not exist without its embodiment in a state of mind. As to the novel's construction, for Woolf, all *interpretations* are the effect of a mental process based on personal background. If this is true, it follows that life is a process made up of transitory unities connected in a *continuum* by a conscious relationship between object and subject.[6]

Septimus commits suicide and Clarissa, coming to know of it, accepts her own life with a renewed joy.

6   In *Tristram Shandy* reality was interpreted in different ways according to the single individual's hobby-horse. However, in *Mrs Dalloway*, we do not find obsessive metaphors. We find only that the traumas and difficult choices made in the past are still conditioning the present.

In *Mrs Dalloway* inner and outer worlds are defined by an individual experience of time and symbolism (the green dress, the roots of a tree, Big Ben, the flowers, the letters in the sky, and so on). In *Orlando*, the relation between inner and outer is mediated by clothes and poetry. On various occasions, Orlando claims that a dress/suit works like the skin since it connects and influences the mind. Orlando behaves differently depending on what s/he is wearing, to the point that his/her clothes modify his/her features, posture and looks (Woolf 1928/1969, p. 132).[7] After changing sex and becoming a woman, she needs to establish contact with the world as a woman — that is, through a skirt (*ibid.*, pp. 108–9). Her difficulties in wearing this article of clothing become a *material* mediation between herself and the real world.[8] Being a woman is not only a cultural commitment, but also a *state of being* mediated by body and appearance.

Orlando loves poetry and s/he often makes use of metaphorical images to decode reality. These images are fixed in her/his and our imagination as an extension of the surroundings, as a way of going beyond appearance to achieve knowledge. When Orlando lives this awareness, s/he 'mislays' her/his senses and enters into a metaphysical dimension. But, paradoxically, the images are anchors dropped by her/him to stay in touch with the world since s/he is essentially a human being living through an unlimited temporal and spatial dimension beyond a simple identitarian condition. Through the images[9] s/he can hold together all her multiple historical identities.

We can point out that, faced with a community strongly conditioned by rigid social conventions, inner life is an exceptional space for dynamism and creativity where intelligence and intuition may advance side by side. It is only in this particular intimacy that a human being can find those *moments of being* (roughly equivalent to Joyce's *epiphanies*), as Woolf calls them, through which the order of reality corresponds to self-awareness. Indeed, for both Orlando and Mrs Dalloway, any state of consciousness is a unique moment in which the subject feels total identity with time and space, along with the mindfulness of her mind (Varela *et al.* 1991).

Moreover, we recognize symbols used by the narrator to connect one consciousness to the other.

7 It would be interesting to delve into the role of the family house Orlando inherits, in order to understand the relationship between the character and the environment over the centuries.

8 While Gwendolen is a perfectly suitable woman for society, Orlando is a woman in-*process*, committed to making a cultural and identitarian subject of herself. Tristram, the other character-in-progress, learns how to be a man through a rational and emotional process. On the contrary, as Orlando is an adult, she needs to adapt herself to the new possibilities (affordances) offered by her new sexual identity and the historical context. As she herself points out, she feels that throughout her life, she has always had the same melancholic nature, in love with death, passion, poetry, landscapes, and animals.

9 According to David Lodge, literature renders *qualia* through metaphor and simile so that one sensation is invoked to give specificity to another. The non-verbal is verbalized. See Lodge (2002).

To a certain extent this is what happens to the reader who embodies a universe of words and images which belong to a fictitious plot culturally and linguistically connected to her own world. The reader comes into existence by means of the reading process: the eye follows a succession of printed words on the page, the ear is virtually solicited by the voices of the characters, verbal tropes stimulate a series of visual experiences and so on. The text brings physically into existence a world of relations and processes much more extended than those physically occurring between the printed page and the body of the reader. Somehow, the reader becomes a huge network of extended processes (Manzotti 2006). Once again, literature seems to tell us that the complex verbal-visual system of signification not only allows the reader to create a particular literary context and to 'give life to' other people's adventures, but it also allows the reader to experience herself. Literature enlarges the reader's life. All literary texts are immersive in their own way.

4 From postmodern to digital literature: From hetero-cosm to virtual narrative environment

Literature moved a step forward in the second half of the twentieth century when poststructuralism and postmodernism revealed that reality is no longer merely reflected by language but *it is* language. Therefore, the notion of consensus (that is the idea of Truth or the existence of a meaning) becomes an illusory social product and is replaced by a network of discourses. All human beings become writer and reader, encoder and decoder of multiple realities conceived as a text to be interpreted, just like a literary text. In literature, the reader is *explicitly* asked to be aware of the creation of meaning through metafictional and metacritical comments. The reader becomes *explicitly* responsible for the enactment of the storytelling, as Laurence Sterne had already theorized in the eighteenth century.

Consider John Fowles' *The French Lieutenant's Woman*. The reader is first asked to participate in the re-construction of a Victorian England setting, as normally occurs in realist nineteenth century novels. Yet, in the famous thirteenth chapter, it is revealed that the story is invented and is a product of imagination. The Victorian context was made real thanks to multiple literary and linguistic conventions which have influenced the reader's mind. Similarly, the main character of the novel—Sarah Woodruff—is *narrated* by other characters and *tells* stories about herself. However, these stories are incoherent and none of them correspond to any single truth about the woman. The reader never knows what has really happened to her. All in all, by means of this interplay of stories, the mysterious character is created. The humanist notion of a unitary and autonomous subject is first established and then subverted. This is a nice literary expression of what Lacan referred to as the de-centring of subjectivity (Lacan 2006).

As Jacques Derrida claimed (Derrida 1998), the subject is *situated*, that is to say it is an ideological construct (depending on differences of race, sex, gender, class and so on) in a cultural system. All characters are a niche of expectations solicited by the information they stimulate both in the co-protagonists of story and in the reader's mind. Postmodernism forces the reader to organize and interpret a subjectivity which does not *belong to* so much as it is *acknowledged as* (technically, *misread*). The character becomes an inter-textual product of the affordances suggested to the observer: all subjects emerge from an interpretative reading/writing experience. There are no individual minds unrelated to a network of texts (cultural and linguistic constructs), and there are no conscious minds without a reader. The way in which a subject interacts with the world is much less important than the way it is over and over misread. Characters exist as an *effect* of encoding and decoding processes of linguistic and cultural inputs and, consequently, the reason for their existence is to be found in the interpreters' mind. A character can be conceived as an interpreter's extension with a transitory meaning. The enactment of the reader corresponds to that of the protagonists: the entire world is actualized by interpretations since the subject emerges as an epistemological process. So, enacting experience is made of active encounters (Noë 2009, p. 473).

It is, however, in digital literature and art that a beholder takes on explicitly the role of producer, author, user, and medium of the stories told. As I have observed elsewhere (Carbone 2002), digital fiction is a form of writing conceived to be read sequentially but not linearly on the screen of a multimedia digital device by an interactive reader. The act of decoding a verbal text is amplified by the visual and motor-sensory stimulation undergone by the reader.

Under the influence of postmodern metafiction, the 'hyper-reader' is asked to become a *performer*. The hyper-reader ought to have a total perception of the immersive digital environment. She is encouraged to bring text to life to its full potentiality through a physical as much as an aesthetic and intellectual experience. The work lives as an extension of the performer who needs to understand how to behave in virtual space by means of her skilful bodily activity. The reader of a book knows that he needs to go from the first to the last page of the volume. The reader of a cyber- or hyper-text or the beholder of an installation is at the start engaged in understanding what she is supposed to do with them— software, applications, purpose and reading processes are always different. On the one hand, the 'best' possible *reading performance* allows the work to express its aesthetic value and, on the other hand, the enactive behaviour brings about changes in the performer's stimulation and, consequently, in her knowledge of the work (Noë 2004, p. 8).

Observing several performers, it is apparent that, although the perception of the virtual artistic space is intuitive, the individual reaction as a cognitive space might pose a few problems. It is very often difficult to grasp immediately the best line of action since users have different levels of understanding or reacting skills. Realist literature relies on predictability in order to encourage the reader to identify herself in the narration. Conversely, digital writing plays with the sensual surprise engendered by multimediality and multimodality so as to foster the beholder's creative powers. It is the creativity of the performer that makes the beauty of the work of art emerge.

An even more particular case is that of the cave: experience in virtual space carries great emotional impact in so far as the performer has the advantage of not being constrained by the physical laws of nature. For instance, she can fly over a roof or walk through a camel without being hurt. However, it is often very difficult to make full use of this unique opportunity since our mind is strongly conditioned by our cognitive habits even when these may be set aside.

Virtual cave is a new frontier of literature since the 'reader' is physically inside a 'virtual room' and the text moves all around her: experience and existence are identical for the user as long as she remains in that specific context. Here, the body of the user and her rational thoughts, along with her perceptive system are crucial. Alva Noë points out that:

> The world makes itself available to the perceiver through physical movement and interaction… [P]erceptual experience acquires content thanks to our possession of bodily skills. *What we perceive* is determined by *what we do* (or what we know how to do); it is determined by what we are *ready* to do… [We] enact our perceptual experience; we act it out. (Noë 2004, p. 1)

Emotional and cognitive experiences in the cave are both spatial and temporal, linguistic and audio-visual. The performer needs to understand how to enact the experience, how to behave according to what is happening, what the work is about, and how communicative processes are being implemented. The quality of perceptual consciousness is not just a neural function caused by and realized in the brain, but it is rather to be considered in terms of patterns and structures of skilful activity akin to a mesh of sensorimotor contingencies: the brain, the world and the body work together to make consciousness happen (Noë 2004, pp. 227–8).

Screen (2002) by Noah Wardrip-Fruin, Robert Coover, Joshua J. Carroll, Shawn Greenlee and Andrew McClain is a very interesting example of cave literary work. The story tells the lives of a married couple and their son. The thoughts of each character are typed onto the walls of the room, while music plays and a synthetic voice reads out what is being typed, until the walls are all crammed full of words. At this point the performer feels trapped in a claustrophobic space similar to the one lived in by the protagonists who see their coexistence as unbearable. Words start falling down onto the floor and the performer realises that she or he can stick them back onto the wall with a joystick she or he holds in her or his hand. At first the rhythm is tolerable and the sticking back into place possible, but as time passes the pace becomes faster and faster and the performer cannot accomplish her task so she instinctively gives up and goes back to her traditional function of reading what is still on the walls. Like the text and the experience, the life of the family also disintegrates. The only things left in the room are silence and immobility. A heap of broken words lie on the floor and the literary experience is over.

It is clear that the evanescence of the narrative framework which has traditionally separated fiction and reality is even more complex and still in search of a definition. Instead of the *sense of reality* sought by realist writers (mimesis of the product), virtual reality and immersive environments try to foster a *sense of presence* through an induced transparency and a suspension of disbelief. The artist tries to make the user unaware of the medium by stimulating a motor-sensory experience based on an intuitive-holistic, renewed approach which combines content, form and environment.

Cave art allows the reader-performer to go beyond mimeses or simulation as it constitutes a physical, intellectual, involving experience which shelives through and which is only possible because a certain environment and expectations have been met. Once again, literary experience shows that the role of the reader is fundamental to reaching aesthetic knowledge, but at the same time it highlights how the medium can condition the act of reading and, consequently, the consciousness of it. Literature is no longer conceived either in the text or in the beholder's inner mental dimension only. Literature is a process enacted by multiple subjects such as the reader and the author (Noë 2004; O'Regan and Noë 2001) and likely to be extended to a mesh of multiple physical processes extended in time and space (Manzotti 2006). To perceive the world is to co-perceive oneself. The awareness of the world and of one's complementary relations to the world are not separable in cave literature (Gibson 1979, p. 141).

5 Conclusions

In the present analysis, I have chosen not to adopt a single theoretical externalist point of view, preferring, on the contrary, to see how different literary poetics have, through the ages, dealt with the relationship between mind and body, between subject and object. The seeds of externalism have always been present in the representation of the world, even if conflicting philosophical perspectives have encouraged to persist in adopting an internalist standpoint. As I tried to show, Laurence Sterne, George Eliot and Virginia Woolf transformed a mental process into a narrative form, so that variations in object-subject interaction set up different ways of conceiving literary communication.

All authors put experience at the core of their story-telling. Sterne conceives experience as a mental process which aims at the construction of multiple coexisting personal worlds. All characters have peculiar, obsessive interpretative keys for reality and they use them as a cognitive tool as well as a medium of communication. Everything in life leads the individual from her hobby-horse back to her hobby-horse. Every individual lives in a private world of meaning and of actual facts. The reader's responsibility is to keep all the different coexisting consciousnesses together in a performance that involves commitment, sense of humour, visual and verbal interpretative skills.

Differently, George Eliot conceives the body (or the 'fibres' under the skin) as a privileged medium between subject and object. The body and the sensory stimuli from the world make the character reacts and, in so doing, the character creates the plot and sparks off a chain of events. Experience is made of actions and reactions to such sensory stimulus information which turn into knowledge once they are verbally elaborated. Experience is embodied.

In Virginia Woolf's world, things *are* experience since nothing exists outside of the characters' consciousness. It implies that reality *is* consciousness — something akin to a succession of convergent and divergent overlapping epiphanies that different human beings individually succeed in recognize. Only by means of these particular moments, a mindfulness of mind and world is achieved.

As I have been at pains to demonstrate, the role of the reader and her personal environment are fundamental to enact a literary text, even when different levels of involvement are required. The reader's physical and cognitive world is extended to include to a certain extent the world imagined by the author (the first reader of the work) into a unified web of relations and actual extended processes.

It has been widely discussed how authors assume an *implicit reader* as a potential audience. Over the past few decades, postmodernism and digital literature asked the reader for an authorial collaboration in the active generation of the work of art itself and not only in the reconstruction of the meaning. So far, literary experience has been determined by the power of attracting our imagination and of overcoming our biological, physical, social, cultural, and ideological boundaries so as to extend our selves into a larger world. However, the most recent experimental literary production demands the beholder to be the author of the literary work often not less than the author; in turn, the author is the first reader. In postmodernist literature, the reader receives linguistic and cultural stimuli from the novel that underpin and endorse the creation of stories about fictitious characters and settings—truths only transitory and partial. Finally, in the cave, the reader's existence and creative performance is coextensive with experience itself. The brain is not enough and literature bears witness to this.

References

Bennett, M.R. and P.M.S. Hacker (2003), *Philosophical Foundations of Neuroscience*, Malden (MA), Blackwell.
Carbone, P., Ed. (2002), *eLiterature in ePublishing*, Milan, Mimesis.
Carbone, P. (2008), *La Lanterna Magica di Tristram Shandy: Visualità e Informazione, Ordine ed Entropia, Paradossi e Trompe L'Oeil nel Romanzo di Laurence Sterne*, Verona, ombre corte.
Davis, M. (2006), *George Eliot and Nineteenth-Century Psychology: Exploring the Unmapped Country. The Nineteenth Century*, Aldershot and Burlington (VT), Ashgate.
Derrida, J. (1998), *Of Grammatology*, Baltimore (MD), Johns Hopkins University Press.
Eliot, G. (1876/1970), *Daniel Deronda*, London, Panther.
Fowles, J. (1969), *The French Lieutenant's Woman*, London, Cape.
Gibson, J.J. (1979), *The Ecological Approach to Visual Perception*, Dallas (TX) and London, Houghton Mifflin.
Honderich, T. (1998), 'Consciousness as Existence', *Royal Institute of Philosophy Supplement*, **43**: 137-99.
Huxley, T.H. (1869/2000), *Methods and Results*, London, Elibron Classics.
James, H. (1884/1948), *The Art of Fiction, and Other Essays*, New York, Oxford University Press.
Lacan, J. (2006), *Écrits: The First Complete Edition in English*, New York, W.W. Norton and Co.
Lewis, P. (2007), *The Cambridge Introduction to Modernism*, Cambridge, Cambridge University Press.
Locke, J. (1690), *Essays on Human Understanding*, Book IV, London.
Lodge, D. (2002), *Consciousness and the Novel: Connected Essays*, Cambridge (MA), Harvard University Press.
Manzotti, R. (2006), 'Consciousness and Existence as a Process', *Mind & Matter*, **4** (1): 7-43.
Noë, A. (2004), *Action in Perception*, Cambridge (MA), MIT Press.
Noë, A. (2009), *Out of Our Heads: Why You Are Not Your Brain*, Cambridge (MA), MIT Press.
O'Regan, K.J. and A. Noë (2001), 'A Sensorimotor Account of Vision and Visual Consciousness', *Behavioral and Brain Sciences*, **24**: 939-73.
Raphael, L.S. (2001), *Narrative Scepticism: Moral Agency and Representations of Consciousness in Fiction*, Madison (NJ), Fairleigh Dickinson University Press.
Sterne, L. (1759), *The Life and Opinions of Tristram Shandy, Gentleman*, London, Ann Ward.
Varela, F.J., E. Thompson and E. Rosch (1991), *The Embodied Mind: Cognitive Science and Human Experience*, Cambridge (MA), MIT Press.
Woolf, V. (1928/1969), *Orlando: A Biography*, Harmondsworth, Penguin Modern Classics.
Woolf, V. (1925/2002), 'Modern Fiction', *The Common Reader*, First Series, Harvest/HBJ Book: 146-54.
Wardrip-Fruin, N., R. Coover, J.J. Carroll, S. Greenlee and A. McClain (2002), *Screen*.

Part Three
Art Beyond the Skin

Teed Rockwell

Musical Experience and the Extended Self

C.S. Lewis pointed out that we have two almost contradictory meanings for the word 'really', which he illustrated with the following sentences: 1) 'All that *really* happened was that you heard some music in a lighted building', and 2) ' It's all very well discussing that high dive as you sit here in an armchair, but wait until you get up there and see what it's *really* like' (Lewis 1942, p. 168). Each of these sentences describes a memorable experience, but each also presupposes a different idea about the nature of experience. The first sentence is intended as a deflation or dismissal of a mystical or religious experience. This deflation is accomplished by narrowing the concept of experience to an awareness of the facts revealed by the sense organs. In the second sentence, experience is seen as a unified whole that includes both this sensory awareness and the emotions or 'feelings' it inspires. In ordinary language we often use similar words to refer to both kinds of experiences. We refer to pain as a feeling, but we use the word 'pain' to refer both to the physical sensations caused by tissue damage, and the emotional reaction we have to those sensations. We also use the word 'pain' to refer to emotional states that have no direct connection to an immediately present sensation, such as the feelings caused by the loss of a job or a friend. We say we feel both sadness and physical warmth, and, in different contexts we apply the word 'real' to both of these very different sorts of feelings.

The experiences referred to in Lewis's first sentence are thought of as the subject matter of the sciences. The experiences referred to in his second sentence are the province of literature and the arts. Let us then refer to experiences of the first type as *sensations* and experiences of the second type as *aesthetic experiences*. It is the artist's job to create this second kind of experience for general consumption, but we also each get it directly from lived life as we encounter sunsets, life challenges, etc. The above examples show that ordinary language presupposes that both of these kinds of experiences are 'real', in some sense. However, our materialist presuppositions about mind and knowledge inevitably incline us towards the assumption that sensations provide contact with the outside objective world, and aesthetic experiences are mere inner subjective reactions.

One presupposition that requires us to devalue aesthetic experience in this way is the mind/brain identity theory. If the mind is identical to the brain, obviously we can only be aware of things that somehow manage to fit inside the brain. Sensations seem capable of working their way into our brains in this manner, when they are conceived of as discrete 'nows', each of which is produced by the immediate contact of a stimulus with a sense organ. However, any structure that emerges from these discrete sensations has to be superimposed by the mind, which means all such structure is internal and subjective, rather than external and objective. We think of the mind/brain as a box, and the sensations as separate bits of information pushed into the box one at a time. Unless the mind/brain gives some kind of order to this jumble of separated bits, it will be chaotic and absurd. This is basically the relationship that allegedly exists between 1) percepts and concepts, and 2) facts and theories. The percepts/facts are the unrelated bits which are given form and coherence by the concepts/theories.

If scientific common sense were completely consistent, it would accept Kant's claim that all higher level structure, both scientific and aesthetic, has to be superimposed by the mind, and that atomistic percepts and facts by themselves tells us nothing about the mind-independent world. In practice, scientific common sense assumes that, at the purely physical level, structural concepts such as gravity, electromagnetism, capacitance, etc. magically emerge from the directly perceived facts. If you take care of the scientific facts, somehow the theories will take care of themselves. However, only physical facts are privileged with this ability, which means that the kind of structure that makes aesthetic experience meaningful has to be superimposed by the experiencing subject. We are thus lumbered with what philosophers call the *fact/value distinction*. Assertions like 'this is red' or 'this weighs five pounds' are about facts and thus describe the external world. Assertions like 'this painting is beautiful' or 'this action is cruel' are statements about value, and describe our feelings about the world, and not the world itself. The unity that is present in aesthetic experience is thus divided into objective and subjective components, with the artist having dominion over the subjective and the scientist having dominion over the objective. So much for Keats' claim that beauty is truth, and truth beauty.

And yet we often speak of art as giving meaning to our lives, just as we speak of a meaningful job, or a meaningful relationship. But 'meaning' in this more overarching sense has no place in a worldview that accepts mind/brain identity theory. I once knew an analytic philosopher who was asked 'what is the meaning of life?' He replied that this question contained a category mistake, because life is a biological process, and meaning is a property that only applies to language. One can coherently ask about the meaning of the word 'life', but not about the meaning of life itself. How then do we get to more complex meanings from the meanings of individual words? According to orthodox Chomskian linguistics, the meanings of individual words can be combined according to the rules of grammar, to create the higher level meanings of sentences, paragraphs and novels. But all of these higher level meanings have to be constructed from their smaller parts using combinatorial rules. George Lakoff lists three presuppositions that he claims, I think correctly, are part of the Chomskian paradigm for linguistics:

(A) Every concept is either a primitive or built up out of primitives by fully productive principles of semantic composition.
(B) All internal conceptual structure is the result of the application of fully productive principles of semantic composition.
(C) The concepts with no internal structure are directly meaningful, and only those are. (Lakoff 1987, p. 279)

If we use this combinatorial theory of meaning to account for aesthetic experience, we see aesthetic experience as constructed from sensations by the mind. This could imply one of two importantly different conclusions. An empiricist approach, reminiscent of British philosopher David Hume, would see this constructive process as pretty much a matter of free choice by the mind doing the experiencing. This was the attitude that was presupposed by the twelve-tone music of Arnold Schonberg and his disciples. The composer assumes that each of the twelve notes of the tempered scale is the only musical element that is 'directly meaningful', to metaphorically extrapolate from the above Lakoff quote. The composer is theoretically free to arrange those twelve tones in any order she wishes, using whatever 'grammatical rules' she chooses to construct. This does not mean that twelve-tone music is chaotic, but it does mean that the strict rules it follows are an artefact created by the composer. One of those rules, for example, is that the twelve-tone row chosen for any given composition cannot repeat any of the twelve notes before it has played all of them. But at the time this twelve-tone row is chosen, there is no objective universal reason why one note should follow or precede any other note. If a particular sequence of notes sounds insane to you, that's just your subjective reaction.

In his series of Harvard lectures entitled *The Unanswered Question*, Leonard Bernstein proposed a more rationalist Kantian alternative to the relationship between notes and melodies. He argued that the diatonic scale was an *a priori* structure in our minds that made musical experience possible, in much the same way that Chomsky's *a priori* generative grammar made language possible. Bernstein constructs a Chomskian analogy between musical experience and language with the following parallel charts (Bernstein 1976, p. 85):

Language	Music
D) Supersurface Structure	D) Surface Structure
(poetry)	(music)
C) Surface Structure	C) Deep Structure
(prose)	('prose')
B) Underlying Strings	B) Underlying Strings
(Deep Structure)	
A) Chosen Elements	A) Chosen Elements

These charts describe a hierarchical relationship of wholes and parts, with the simplest elements at the bottom, and their composites becoming progressively more complex as we move up the charts from A to D. The phrase 'chosen elements' at the bottom of each side of the chart refers to a range of assemblages. At the bottom of the language chart are those perceptual sensations called phonemes, which are assembled into morphemes, which are in turn assembled into words. The words are in turn assembled into B) underlying strings, which constitute the deep structure that encodes the fundamental meanings of our linguistic utterances. The utterances themselves, however, have C) surface structure, which transforms the deep structure by deleting and rearranging various elements of it. This creates a kind of ambiguity which streamlines communication for common sense social purposes. The goal of the analytical work done in linguistics is to uncover the relationship between the ambiguous surface structure of ordinary speech and the deep structure which most accurately conveys its meaning. There is also a higher form of communication that uses even more deletion and transformation to create the greater ambiguity of what Bernstein calls the A) 'super surface structure' of poetry.

Bernstein creates a musical analogy with Chomsky's concepts by seeing the diatonic scale as roughly analogous to the phonemes and morphemes of spoken language. By themselves individual notes don't say anything musically, just as an individual letter has no semantic content. The same note could be part of the Star Spangled Banner, Eleanor Rigby, or any other imaginable melody, depending on what context it occurs. Notes acquire meaning when they are heard as parts of riffs or motifs, which in turn acquire their higher level meanings as parts of longer melodies, which in turn acquire still higher level meaning as parts of symphonies or operas. Bernstein admits that the analogy between melodies and words is not precise (at one point he says that the musical equivalent of a sentence would have to be an entire symphonic movement: Bernstein 1976, p.61; Bernstein 1992, disc 2).[1] The important point for our purposes, however, is that both music and language assemble meaningless parts into meaningful wholes by means of innate structural rules.

Bernstein assumes that we must choose between 1) innate musical structures in the mind, and 2) music as independent notes structured only by subjective choice. His arguments and examples eloquently illustrate that the first choice is definitely preferable. But are these our only choices? Yes, if we refuse to question the metaphor of the mind/brain as a box into which one dumps notes, which are then assembled by the mind. Suppose the mind is not a box? Suppose we do not have to break the world up into bite-size chunks and cram them into the brain before we can be aware of the world? Rejecting this assumption would make the first step in Bernstein's charts unnecessary. Bernstein must start with a foundation of 'chosen elements', and the task of assembling those elements into a deep structure of underlying strings, only because it is assumed that this is the only way that we can ever get anything into the brainbox. However, there is an alternative way of seeing the growth of knowledge which makes this step unnecessary. In this chapter I will refer to this idea as the Hypothesis of the Extended Self, or HES. This is a deliberate reference to the acronym HEC for the Hypothesis of Extended Cognition, which is now used to refer to several closely related cognitive theories (Rockwell 2005; Wilson 2004; Clark 2008; etc.) HES, however, is committed to a no doubt controversial explanation for why cognition is extended: that we begin our conscious cognitive lives not isolated from the world, but unified with it.

1 The Self and the Other

Freud's description of an infant's pre-epistemic relationship with the world is radically different from the presupposition shared by both Hume and Kant, i.e. that we start out separated from the world, and that the goal of knowledge acquisition is to bridge this gap.

> An infant at the breast does not as yet distinguish his ego from the external world as the source of the sensations flowing in upon him. He gradually learns to do so, in response to various promptings... in this way there is for the first time set over against the ego an object, in the form of something which exists 'outside'... One comes to learn a procedure by which, through a deliberate direction of one's sensory activities and through suitable muscular action, one can differentiate between what is internal — what belongs to the ego — and what is external — and what emanates from the outer world... In this way, then, the ego detaches itself from the external world. *Or to put it more correctly, originally the ego includes everything, later it separates off an external world from itself.* Our present ego feeling is therefore only a shrunken residue

---

[1] It is very difficult to follow Bernstein's arguments without hearing his musical examples. For that reason, I would recommend using both the DVD and book versions of these lectures.

of a much more inclusive—indeed an all embracing—feeling which corresponded to a more intimate bond between the ego and the world about it. (Freud 1930/1961, p. 13–5, italics added)

The removal of certain passages in the above quote makes it easier to isolate the epistemological implications from the psychiatric and theological issues that primarily interested Freud. An infant learns not by acquiring and assembling bits of information, but by dividing a unified range of experience into more and more parts of varying sizes. When the infant first discovers that the mother's breast is not always available, its world is divided into Self and Other for the first time. As time goes on, the infant divides the Other into a variety of other categories—mother, father, bed, toys etc. until eventually the child is a mature adult with a sophisticated category system. If she is a Humean, the adult then assumes that the only way she could possibility understand this multi-faceted world is to analyse it into attributes and qualities, and then make copies of those attributes inside her brain. If she is a Kantian, she makes the further inference that she must reassemble those attributes into a more comprehensive representation by means of *a priori* cognitive rules. Neither of these steps is necessary if our experiences were carved out of a primordial whole, rather than assembled from fundamental bits. The carving process creates categories and objects of a variety of sizes, and all of them have an equally close relationship to the primordial whole. This view of knowledge frees us from the assumption that all of our experiences must be assembled from what David Hume called simple ideas (Hume 1888/1964, p. 3). Because the introspectionist psychology of Edward Titchener failed spectacularly in its search for these simple ideas (Guzeldere 1995) this is an assumption we are well rid of.

One might argue that our experience only *appeared* to be unified with our environment during our infantile innocence, and that this innocence was a form of ignorance, not awareness. Freud was inclined to think this himself, but he took very seriously Romain Rolland's suggestion that sustaining this 'oceanic feeling' was the goal of religious and meditative practices. Rolland thought this feeling was an awareness of our fundamental oneness with the universe, but Freud, who admitted he had never experienced such a feeling himself, saw it as 'something rather in the nature of an intellectual perception' and remarked that 'I could not convince myself of the primary nature of such a feeling' (Freud 1930/1961, p. 8). What I will be arguing, however, is that Rolland was essentially right. This experience of oneness is primary, and provides the starting point for the diversity that emerges from analytical activities. This is why aesthetic experience is more fundamental than sensations, which are an artefact created by our assumption that experience must be divided into subjective and objective components. In much the same way, we divide our experience of speech-acts-performed-in-lived-social-contexts into individual sentences, which we in turn analyse into individual words. We also analyse our experience of music at concerts or dances into individual melodies, which we then analyse into individual notes. However, none of this implies that we can only *have* the experience by copying regions of the outside world small enough to fit inside the head, then assembling them into a whole using *a priori* rules. The experience begins as a whole, and thus (as James first said) the act of analysis is the achievement, not the act of synthesis.

As unfashionable as this idea may seem at the moment, it has had a variety of distinguished defenders. Both Dewey and Heidegger believed that this harmonious identity[2] between self and world was present not just in moments of 'oceanic' mysticism, but in everyday interactions in which our tools and practices are what Heidegger calls 'ready-to-hand'.

> In well-formed, smooth running functions of any sort—skating, conversing, hearing music, enjoying a landscape—there is no consciousness of separation of the method of the person and of the subject matter. When we reflect upon experience instead of just having it, we inevitably distinguish between our own attitude and the objects towards which we sustain that attitude… such reflection upon experience gives rise to a distinction of what we experience (the experienced) and the experiencing—the how. This distinction is so natural and so important for certain purposes, that we are only too apt to regard it as a separation in existence and not as a distinction in thought. Then we make a division between a self and the environment or world. (Dewey 1916, pp. 166–7)

This blurring between self and world was also implied by William James's theory of radical empiricism, which saw the relations that connect experience to be as real as the items that were connected. James realized that once relations were seen to be as fundamental as relata, the line between the two would no longer be sharp. He does say that 'Empiricism… is essentially a mosaic philosophy' (James 1912/2003, p. 22), but he then turns that metaphor on its head by adding 'in radical empiricism, there is no bedding, it is as if the pieces clung together by their edges, the transitions experienced between them forming their cement… there is in general no separateness needing to be overcome by an external cement' (*ibid.*, p. 45). This does not mean that all experience is an homogenized undifferentiated mush, but it does mean that individual experiences are more like waves on an ocean than rocks in a pile: noticeable features of a single process, rather than individual items clustered together. Once James acknowledged that experience was not broken into fundamental parts, he naturally concluded that

---

2 I realize that the idea of 'harmonious identity between' is arguably a contradiction in terms. Identity is not a relationship between two things, it is a relationship that something has with itself (which is arguably also a contradiction). All of the writers quoted above alternate between claiming that the mind and the world are one, and then talking about them as if they were two interrelated but distinct items. However, this tension is inevitable when one is trying to replace one set of concepts with another. We can't say 'that table is actually an amalgamated structure of molecules' without assuming that it is in some (perhaps illusory) sense also a table.

there was also no sharp line between experience and the world. He objects to the fact that 'Throughout the history of philosophy, the subject and its object have been treated as absolutely discontinuous entities' and says that philosophy should instead 'discuss the ways in which one experience may function as the knower of another' (*ibid.*, p. 28).

To add a final very strange bedfellow to this group, logical positivist philosopher and physicist Ernst Mach asserts that 'The ego is not a definite, unalterable, sharply bound unity' and 'the ego can be so extended as ultimately to embrace the entire world' (Mach 1886/1959). Mach was fundamentally a Humean atomist, but he was an atomist both ontologically and epistemologically. He believed not only that our experience was divided into directly perceived distinct elements, but also that these elements were the fundamental constituents of reality itself. Precisely for that reason, however, he felt that any other way of dividing reality was pragmatic and provisional, including the psychological border between subject and object: 'The supposed unities "bodies" and "ego" are only makeshifts... we find ourselves obliged, in many more advanced scientific investigations, to abandon them as insufficient and appropriate' (*ibid.*). Consequently, although he was a physical atomist, he was a psychological holist. He describes his philosophy coming to him in a mystical flash of insight that strongly resembles Freud's and Rolland's oceanic feeling.

> On a bright summer day in the open air, the world with my ego suddenly appeared to me as *one* coherent mass of sensations, only more strongly coherent in the ego... we must not allow ourselves to be impeded by such abridgments and delimitations as body, ego, matter, spirit, etc. which have been formed for special, practical purposes and with wholly provisional and limited ends in view. (*ibid.*)

Although citing such eminent and diverse names may partially mute the sense that this idea is crazy or unrespectable, we'll need actual arguments to provide support for the claim that it is true. The basic arguments I will rely on are 1) The Hypothesis of an Extended Self (HES) is fully compatible with data that is usually interpreted as supporting the brainbox theory of mind; 2) The brainbox theory has problems of its own which the HES effectively bypasses, including problems about the downgrading and subjectivizing of aesthetic experience; 3) The HES has strong support from Connectionist Neural Network theory, which is our most biologically plausible account of how the mind works. Many of my arguments will be based on anecdotes and personal observations, others will be based on what I think are plausible speculations from recent scientific data. New ideas have to start somewhere, and personal experience and hunches are the usual starting points for both scientific and artistic insights.

2 Bernstein's theory without the bottom layer

Suppose we eliminate Bernstein's bottom layer of 'chosen elements', i.e. fundamental distinct words or pitched notes. Suppose we embrace a kind of musico-linguistic string theory, and see the fundamental deep structure as a network of underlying strings, which interact in such a way as to assimilate into each other, and reform into completely new strings. Would this have any impact on the kind of Chomsky-style analysis being done by Bernstein (or for that matter, by Chomsky)? There are people who have argued that Chomskian linguistics doesn't work that well even for language, most prominently George Lakoff. Many of Lakoff's criticisms attack the Chomskian assumption that we perceive and categorize the world by analysing it into some sort of category system. In Chomsky's system, everything we encounter is some sort of X, with clearly defined borders to distinguish it from non-Xes: an apple, a hammer, a traffic violation, a C#. We think about the world by giving a name to each of these categories, and combining and manipulating these names by the rules of grammar and logic: Charles ate the apple, Jack loves Jill. Lakoff argues, however, that our category systems do not divide the world into distinct kinds, but rather cluster items together around a central prototype. The connections between these items radiate out to create a blurry network of similar items related only by what Wittgenstein called family resemblance. There are no 'essential properties' possessed by all members of the category, or sharp lines that cleanly separate one category for another. There are only clusters of properties possessed in different amounts by different members of partially overlapping categories.

If Lakoff is right about this, it would be one reason why Chomskian rules cannot create all meaningful sentences entirely out of individual words and rules of grammar. If the borders of the categories are blurry, they will probably refer to slightly different items depending on the sentences in which they are used. Consider the word 'red' and how its meaning changes when one talks about red hair, red meat, red wine, or red herring. Because the meaning of each individual word changes depending on the sentence, 'grammatical constructions in general are holistic. that is… the meaning of the whole construction is motivated by the meaning of the parts, but is not computable with them' (Lakoff 1987, p. 465). The bottom line of 'meaningful' words apparently doesn't provide the foundation that would enables us to build all meaningful sentences using nothing but grammatical rules. The plausible conclusion from this is that the sentences have meanings of their own, as do the paragraphs they inhabit, as do the novels and essays the paragraphs inhabit, and so on *ad indefinitum*.

It's hard to consider the implications of this claim when we are considering all categories from aardvarks to zymurgy. Perhaps the implications will be clearer if we consider a category system which is confined to only one sensory modality (hearing) and consists of only twelve distinct kinds of perceptible items (the twelve notes of the tempered scale). Need we assume that it is only possible to perceive a melody by perceiving each individual note and then combining those notes into longer strings by the rules of diatonic harmony and modulation? I maintain that our comprehension of music goes in the opposite direction. We perceive entire melodic 'sentences' as wholes. Occasionally when we musicians need to be sure we are playing all aspects of a melody correctly, we break it down into individual notes for fine-tuning purposes. The process is essentially the same process that was started by the infant who divided up a unified world into self and other, then into mother, father, bed etc. However, unless there is a compelling pragmatic reason to be aware of the individual notes, this divisive process does not necessarily travel down that far, only to complete melodies of varying lengths.

I believe this is why learning half of a melody you already know does not come automatically and easily, but is almost like learning a whole new melody. If we had all of the notes stored individually, it should require only a short line of neurological 'code' to insert a break point in the string of discrete notes. Instead, the melody acquires a whole new feel, and has to be learned almost from scratch. The fact that it is not fully as hard as learning a new melody can be accounted for by Lakoff's radial prototype theory. The melody-half resembles the whole melody, and learning new melodies consists of radiating out from the prototypical theme to develop similar variations. In fact, I think that most of Bernstein's evocatively loose musical application of Chomsky's grammatical transformations can be described equally effectively as 'radial categories' that branch out from a central prototype (Lakoff 1987, pp. 83–4). Although I will not here attempt a reinterpretation of Bernstein in Lakofffian terms, the following troubling problems with the Chomskian view give us good reason to take that possibility seriously.

3 Troubles with Chomskyism

There are people who can play music but who cannot identify individual notes because they know no music theory. Some of them are geniuses: Django Rhinehardt, Carlos Santana, Eric Clapton. These people not only have complex muscular skills, they also know how to make very sophisticated decisions as to what sort of music to play when. When she hears other musicians play a blues shuffle, a skilled but musically illiterate musician plays only riffs that fit with a blues shuffle. When the other musicians play a bossa nova, this same musician can often play another very different set of riffs that is appropriate for this other context. And yet if you talk to these people (which you have to do if you're in a band with them) you discover that many of them have verbal skills that are only marginally better than Washoe the Chimp's. Consequently, it seems unlikely that they are doing what they are doing with some kind of rational or verbal processing. Defenders of the Chomskian paradigm often reply to this objection by saying that all of these people who appear to be thinking non-verbally have a language of thought inside their heads, which is disconnected from their speech centres. That's why they can't talk about what they do. But there's no real proof that this language of thought exists, and there are serious technical obstacles that would prevent biological humans from implementing a Chomskian style language of thought system.

After devoting almost half an hour to a linguistic analysis of a short fragment of great music, Bernstein makes this comment:

> In fact when you think of the number of transformations taking place in the short space of those few bars of Brahms, it becomes almost incredible that all of them can be instantaneously perceived. (Bernstein 1976, p. 27)

I don't find this almost incredible; I find it completely unbelievable. Bernstein admits that he's leaving out most of what we hear even with these detailed descriptions. Processing the kind of information he describes might be possible for a modern computer, which is nowhere near as good as a human listener at responding sensitively to music. But computers run their programs at the speed of light, and signals in human brains only travel at about 200 miles an hour. If we ran a computer program of this sort on a human brain, it would run at less than a thousandth of the speed of a silicon computer. So how can a brain do more in less time than a computer can? There has to be a difference in the software to get such good results on such slow hardware.

Another problem is that this Chomskian paradigm implies that all of the information we pick up from music can be captured in something like the western notation system. Those who learned in the European classical tradition (such as Bernstein) usually see their primary skill as producing performances from sheets of paper, which is why the linguistic metaphor is so plausible when applied to symphonies etc. However, there are a great number of parameters in music that can't be captured with any sort of music notation, and vocalists and instrumentalists who don't rely heavily on music notation learn how to manipulate these with tremendous skill. European classical performers usually have trouble hearing these non-transcribable nuances, which is why people who can't play those nuances are often accused of 'sounding white'. These nuances are, however, the essential artistic element of most improvised music. Indian vocalists, for example, use what they call *sruti*; the microtones between the notes that cannot be written on paper. And although Indian written music has some very approximate ways of indicating where those *sruti* are supposed to be played, no one would ever think these marks could tell you exactly how to play or sing those *srutis*. In fact what most Indian music teachers tell their students is 'don't think you've learned this part just because you can play what's on the paper. Forget about what's on the paper and listen to what I'm playing.'

This is equally true for musicians and vocalists in jazz, country, Irish music, and almost any other kind of music that is fundamentally an oral tradition. These oral traditions see the sheet music (when they use it) as marking out the rough outlines of the melody, rather like the key statistical points on a graph. But the melody itself is not enumerated by the written notation system, and its full richness is too complex to be captured in any such system. Fortunately, when one learns the melody as a whole, the parts do not need to be enumerated. Some people object to this idea because it seems to make learning a melody a magical and incomprehensible process. How else could we learn to recognize a melody other than breaking it up into notes? This question can be deflected with another question: how do we learn to recognize individual notes? By breaking them up into parts? How then do we learn to recognize those parts? And so on *ad infinitum*. This problem is thus equally present in the Chomskian system, and is somewhat easier to ignore only because it has been kicked downstairs. This problem disappears however, when we see perception and categorization as a process of dividing a whole, rather than assembling parts. More importantly, Connectionist Neural Nets, which are currently the most biologically plausible way we have of representing human cognition, do operate this way, so we no longer need to think of holistic cognitive processing as an evocative mystery.

4 Why connectionist networks are holistic

The basic science of Connectionist Neural Networks has been explained in many places (Bechtel and Abrahamsen 1991; Churchland 1995; Rockwell 2005, pp. 183–92), and the philosophical implications of this science have been extensively discussed. I will assume that my readers are at least somewhat familiar with this science (or are willing to become so by checking the above references).

A connectionist net operates by taking an array of input values and transforming it into an array of output values. In an organism with a nervous system, the input values are neurological signals from sense organs and the output values are neurological signals sent to muscles etc. Because an array of numerical values is called a *vector* by mathematicians, this kind of cognitive processing is known as *vector transformation*. Vector transformation operates by rules that are fundamentally different from linguistic processing. There are no discrete words or sensations to be manipulated, no items that have to be moved from one address to another address. Instead, there is a correlation between kinds of input and kinds of outputs.

This correlation is facilitated by intermediate layers of neurons that transform the input signals before sending them on to the final output layer.

Perhaps the simplest vector transformation machine is a thermostat. Its input 'vector' is a single value (the temperature) and its output 'vector'[3] is whatever makes the furnace increase its temperature, e.g. the amount of electricity going to a heating coil. This system performs a useful function by varying one parameter proportionally to the other, but it is so simple that very few people would call it cognitive. With a full-blown connectionist system, however, we get several distinct kinds of outputs in response to several distinct kinds of inputs. This is what enables connectionist systems to function as categorical classifiers. The categories they create are family resemblances that cluster around prototypes, rather than Aristotelian Categories which share the same essential properties. Nevertheless, they do divide up the world into comprehensible regions, and like Freud's hypothetical infant, they become more cognitively sophisticated by starting with a unified whole, and then dividing the world into smaller and more complex parts. One of the simplest of such connectionist networks, discussed in Churchland (1995), processes the inputs from a submarine sonar system, and divides what it perceives into two categories: mines and rocks. The trainers of this system designate one set of values of the final output layer as signifying 'mines' and the other as signifying 'rocks', then tunes the intermediate layers until the final outplayer layer responds with the 'rocks' signal when the input layer encounters a rock, and with the 'mines' signal when the input layer encounters a mine. As far as this system is concerned there are only two entities in the world, mines and rocks, and everything it encounters either fits into one of these categories or doesn't register cognitively at all. More sophisticated connectionist systems divide the world into a greater variety of categories. Typefaces or handwriting can be classified into the twenty-six letters of the alphabet, regardless of the type font or penmanship. Audially responsive systems can tell the difference between the words 'departure' and 'arrival', regardless of whether they are spoken by a 12-year-old Chinese girl or a 50-year-old Hispanic man. Face recognition systems can not only recognize new photographs of the persons they were trained to recognize, but correctly classify the sex of faces they have never seen before. All of this is done by starting with a unified region of possibility space delineated by the output vectors, which is then divided by fine tuning its relationship with the input vectors. There is thus no reason to assume that we must learn to distinguish one melody from another by enumerating the individual notes in

---

3   Strictly speaking, a 'vector' with only one value is called a scalar. But as I am trying to show the bottom limit of the concept of vector by speaking of a 'vector' with only one value, I'm going to use 'vector' in quotes to show how we progress from the simplest possible vector transformation system to one which is capable of processes that are unambiguously cognitive.

each, then tying them together with scalar and harmonic rules. The Monty Pythons ridiculed this view of language acquisition by imagining an actor who memorized all the individual words of a Shakespeare script and then said 'We've got all the words, now we've got to get them in the right order'. Bernstein's Chomskian view of melody acquisition is basically similar to this, only using notes and melodies, instead of words and sentences.

In actual fact, a neural network would not need to register the individual notes to be able to tell one melody from another. A recurrent neural net with feedback loops can respond to patterns that take place over time (see Churchland 1995, Chapter 5). A network of this sort could recognize two examples of the same melody by giving the same response to both, and could make this response without registering the value of the individual notes. This kind of holistic categorization can be seen in face recognition programs.

> We might have expected each of these cells to become focused on some localized facial feature such as nose length, mouth width, eye separation and so forth. But reconstructing the actual preferred stimuli of the 80 face cells reveals that the network settled into a coding strategy quite different from this... each cell comprehends the *entire surface* of the input layer, and represents an entire facelike structure, rather than an isolated facial feature of some sort or other. (Churchland 1995, p. 47, italics in original)

If it is possible to perceive an entire face without representing isolated facial features, it should be equally possible to perceive a melody without representing isolated individual notes.

5 Dissolving the fact/value distinction: Why aesthetic experience is more fundamental than sensation

If our perceptual system works by dividing up a unified experienced whole, rather than assembling experience from discrete simple elements, this would mean that Lewis's second use of the word 'really' is the accurate one. The experience of jumping from a high dive is not cobbled together out of sensations that were 'really' perceived externally and emotions that were only on the inside. That experience is a unified awareness of the here and now, in which the jump is really scary and the water is really blue. That unified awareness can be divided into sensations and emotions after the fact, but that does not mean that the sensations are objective and the emotions are subjective add-ons. As Freud is sometimes willing to admit, unified awareness is literally our birthright. Although we stray from this infantile Eden by dividing the world into categories of varying breadth and precision, we never completely lose contact with the underlying unity of our experiences. This is especially obvious with sensitive but theoretically illiterate music listeners, who can respond to the overall impact of music while being completely incapable of analysing the music itself. Such people are very different from those who have neither knowledge of nor sensitivity to music. People who are musically sensitive but musically illiterate can tell that music is sad without realizing that it is in a minor key, happy without realizing it is in a major key, and respond intensely to a broad range of other emotional attitudes in music. They can often make accurate judgments of aesthetic quality, both in performance and composition, even though they are incapable of verbally justifying those judgments. I even saw one of my nieces respond appropriately to musical mood when she was less than a month old, and probably still in Freud's pre-individualistic mental state. This kind of response is possible because we can be aware of what music means without being able to analyse it into parts, and because the meaning resides in the music itself, rather than being tacked on afterward by our minds.

References

Bechtel, W. and A. Abrahamsen (1991), *Connectionism and the Mind: An Introduction to Parallel Processing in Networks*, Oxford, Blackwell.
Bernstein, L. (1976), *The Unanswered Question*, Cambridge (MA), Harvard University Press.
Bernstein, L. (1992), *The Unanswered Question* (DVD), West Long Branch (NJ), KULTUR.
Churchland, P.M. (1995), *The Engine of Reason, The Seat of the Soul*, Cambridge (MA), MIT Press.
Clark, A. (2008), *Supersizing the Mind*, Oxford, Oxford University Press.
Dewey, J. (1916), *Democracy and Education*, New York, Macmillan.
Freud, S. (1930/1961), *Civilization and its Discontents*, New York, Norton and Co.
Guzeldere, G. (1995), 'Consciousness: What it is, How to Study it, What to Learn from its History', *Journal of Consciousness Studies*, **2** (1): 30–51.
Hume, D. (1888/1964), *Treatise of Human Nature*, Oxford, Oxford University Press.
James, W. (1912/2003), *Essays in Radical Empiricism*, Mineola (NY), Dover Publications.
Lakoff, G. (1987), *Women, Fire, and Dangerous Things*, Chicago, University of Chicago Press.
Lewis, C.S. (1942), *The Screwtape Letters*, San Francisco, HarperOne.
Mach, E. (1886/1959), *The Analysis of Sensations and the Relation of the Physical to the Psychical* (intro remarks: anti-metaphysical), [Online], http://www.marxists.org/reference/subject/philosophy/works/ge/mach.htm
Rockwell, T. (2005), *Neither Brain nor Ghost: A Nondualist Alternative to the Mind/Brain Identity Theory*, Cambridge (MA), Bradford Books, MIT Press.
Wilson, R. (2004), *Boundaries of the Mind*, Cambridge, Cambridge University Press.

Sylvain Le Groux
& Paul F.M.J. Verschure

Music is All Around Us
A Situated Approach to Interactive Music Composition

Externalism considers the situatedness of the subject as a key ingredient in the construction of experience. In this respect, with the development of novel real-time, real-world expressive and creative technologies, the potential for externalist aesthetic experiences are enhanced. Most research in music perception and cognition has focused on tonal concert music of Western Europe and given birth to formal information-processing models inspired by linguistics (Lerdhal and Jackendoff 1996; Narmour 1990; Meyer 1956). These models do not take into account the situated aspect of music, although recent developments in cognitive sciences and situated robotics have emphasized its fundamental role in the construction of representations in complex systems (Varela *et al.* 1991). Furthermore, although music is widely perceived as the 'language of emotions', and appears to deeply affect emotional, cerebral and physiological states (Sacks 2008), emotional reactions to music are in fact rarely included as a component to music modelling. With the advent of new interactive and sensing technologies, computer-based music systems evolved from sequencers to algorithmic composers, to complex interactive systems which are aware of their environment and can automatically generate music. Consequently, the frontiers between composers, computers and autonomous creative systems have become more and more blurry, and the concepts of musical composition and creativity are being put into a new perspective. The use of sensate synthetic interactive music systems allows for the direct exploration of a situated approach to music composition. Inspired by evidence from situated robotics and neuroscience, we believe that in order to improve our understanding of compositional processes and to foster the expressivity and creativity of musical machines, it is important to take into consideration the principles of parallelism, emergence, embodiment and emotional feedback. We provide an in depth description of the evolution of interactive music systems, and propose a novel situated and interactive approach to music composition. This approach is illustrated by a sensate interactive music system called the SMuSe (Situated Music Server).

1 Computer-based music composition

One of the most widespread computer-aided composition paradigms is probably still that of the music sequencer. This model is somehow a continuation of the classic composition tradition based on the writing of musical scores. Within the sequencer paradigm, the user/composer creates an entire piece by entering notes, durations or audio samples on an electronic score. Due to its digital nature, this score can later be subjected to various digital manipulations. Within the sequencer paradigm, the computer is 'passive', and the composer produces all the musical material by herself. The human is in control of the entire compositional processes and uses the computer as a tool to lay down ideas and speed up specific tasks (copying, pasting, transposing parts, etc.)

In contrast with the standard sequencer paradigm, computer-based algorithmic composition relies on mathematical formalisms that allows the computer to automatically generate musical material, usually without external output. The composer does not specify directly all the parameters of the musical material, but a set of simple rules or input parameters, which will be taken into account by the algorithm to generate musical material. In this paradigm, the computer does most of the detailed work and the composer controls a limited set of initial global parameters.

Some mathematical formulae provide simple sources of quasi-randomness that were already extensively used by composers before the advent of computers. In fact, Fibonacci sequences and the golden ratio have been inspiring many artists (including Debussy, Bartok, Stravinsky, etc.) for a long time, while more recent models such as chaotic generators/attractors, fractals, Brownian noise, and random walks are exploited by computer technologies (Ames 1987).

Different approaches to algorithmic composition inspired by technical advances have been proposed and tested. The main ones are statistical methods, rule-based methods, neural networks and genetic algorithms.

In the wealth of mathematical tools applied to algorithmic composition, Markov chains play a unique role as they are still a very popular model probably thanks to their capacity to model and reproduce the statistics of some aspects of musical style (Ames 1989). Markov-based programs are basically melody-composing programs that choose new notes (states) depending on the previous note (or small set of notes). The Markov state transition probabilities can be entered by hand (equivalent to entering *a priori* rules), or the rules can be extracted from the analysis of statistical properties of existing music (Assayag and Dubnov 2002).

One of the most refined and successful examples of a style modelling system is EMI, Experiments in Musical Intelligence (Cope 1996), that analyses a database of previous pieces for harmonic relationships, hierarchical information, stylistic traits, and other details and manages to generate new music from it.

With the advent of new programming languages, communication standards and sensing technologies, it has now become possible to design complex real-time music systems that can foster rich interactions between humans and machines (Rowe 1993; Winkler 2001; Zicarelli 2002; Wright 2005; Puckette 1996). (Here we understand interaction as 'reciprocal action or influence' as defined in the Oxford New Dictionary of American English—Jewell *et al.* 2001.) The introduction of a perception-action feedback loop in the system allows for real-time 'evaluation' and modulation of the musical output that was missing in more traditional non-interactive paradigms. Nowadays, one can 'easily' build sensate music composition systems able to analyse external sensory inputs in real time and use this information as an ingredient of the composition.

These two aspects (real-time and sensate) are fundamental properties of a new kind of computer-aided composition system where the computer-based algorithmic processes can be modulated by external real-world controls such as gestures, sensory data or even musical input directly. Within this paradigm, the composer/musician is in permanent interaction with the computer via sensors or a musical instrument. The control over the musical output of the system is distributed between the human and the machine.

Emblematic recent examples of complex interactive musician/machine music systems are Pachet's Continuator (Pachet 2006), which explores the concept of reflexive interaction between a musician and the system, or the OMax system based on factor oracles (Assayag *et al.* 2006), that allows a musician to improvise with the system in real time.

2 Interactive music systems

Interactivity has now become a standard feature of multimedia systems that are being used by contemporary artists. As a matter of fact, real-time human/machine interactive music systems have now become omnipresent as both composition and live performance tools. Yet, the term 'interactive music system' is often used for many related but different concepts.

2.1 Taxonomy

Early conceptualization of interactive music systems have been outlined by Rowe and Winkler in their respective books that still serve as key references (Rowe 1993; Winkler 2001).

For Rowe, 'Interactive computer music systems are those whose behavior changes in response to musical input. Such responsiveness allows these systems to participate in live performances of both notated and improvised music' (Rowe 1993).

In this definition, one can note that Rowe only takes into consideration systems that accept musical inputs defined as a 'huge number of shared assumptions and implied rules based on years of collective experience' (Winkler 2001). This is a view founded on standard traditional musical practice. Many examples of augmented or hyperinstruments (Machover and Chung 1989) are based on these premises.

In this context, Rowe provides a useful framework for the discussion and evaluation of interactive music systems (Rowe 1993). He proposes a taxonomy along the three main axes of performance type that ranges from strictly following a score to pure improvisation, musical interaction mode which goes from sequenced events to computer-generated events, and playing mode that illustrates how close to an instrument or a human player the system is.

Score-driven systems rely on predetermined events that are triggered at fixed specific points in time depending on the evolution of the input, whereas performance-driven systems do not have a stored representation of the expected input.

Winkler extends Rowe's definition and proposes four levels of interaction (Winkler 2001). The conductor model, where the interaction mode similar to that of a symphony orchestra, corresponds to a situation where all the instruments are controlled from a single conductor. In the chamber music model, the overall control of the ensemble can be passed from one lead instrument to another as the musical piece evolves. The improvisational model corresponds to a jazz combo situation where all the instruments are in control of the performance and the musical material, while sharing a fixed common global musical structure, and the free improvisation model is like the improvisation model but without a fixed structure to rely on.

Once the musical input to the interactive system is detected and analysed, the musical response can follow three main strategies. Generative methods apply different sets of rules to produce a musical output from some stored original material, whereas sequenced methods use prerecorded fragments of music. Finally, transformative methods apply transformations to the existing or live musical material based on the change of input values.

In the instrument mode, the performance gestures from a human player are analysed and sent to the system. In that case, the system is an extension of the human performer. On the other hand, in the player mode, the system itself has a behaviour of its own, a personality.

2.2 Limitations

The interaction between a human and a system or two systems is a process that includes both control and feedback, where the real-world actions are interpreted into the virtual domain of the system (Bongers 2000). If some parts of the interaction loop are missing (for instance the cognitive level in Figure 1), the system becomes only a reactive (vs. interactive) system. In most of the human/computer musical systems, the human agent interacts whereas the machine reacts. As a matter of fact, although the term interactivity is widely used in the new media arts, most systems are simply reactive systems.

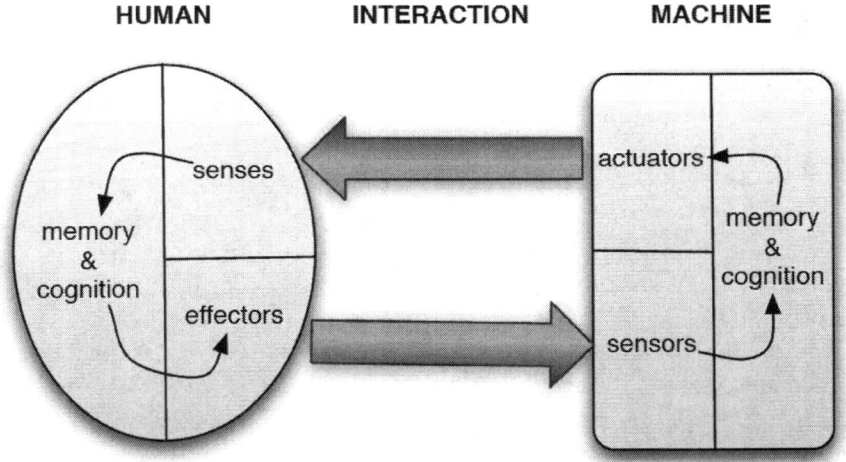

Figure 1: Human machine interaction (adapted from Bongers 2000).

Within Rowe and Winkler's frameworks, the emphasis is put on the interaction between a musician and the interactive music system. The interaction is mediated either via a new musical interface or via a pre-existing musical instrument. This approach is anchored in the history of Western classical music performance. However, with new sensor technology, one can extend the possibilities of traditional instruments by creating new interactive music systems based on novel modes of musical interaction. These systems can generate musical output from inputs which are not necessarily musical (for instance they could be gestures, colours, spatial behaviours, etc.)

The framework proposed by Rowe to analyse and design musical systems relies mainly on what he calls the sensing-processing-response paradigm. This corresponds to what is more commonly called the sense-think-act paradigm in robotics and cognitive science (Pfeifer and Scheier 2001). It is a classical cognitive science approach to modelling artificial systems, where the different modules (e.g. perception, memory, action) are studied separately. Perceptual modules generate symbols representing the world, those symbols are stored in memory and some internal processes use these symbols to plan actions in the external world. This approach has since been challenged by modern cognitive science, which emphasizes the crucial role of the perception-action loop as well as the interaction of the system with its environment (Verschure *et al.* 2003).

3 Designing modern interactive music systems

3.1 A cognitivist perspective

A look at the evolution of our understanding of cognitive systems put in parallel with the evolution of composition practices (which do not necessarily rely on computer technology) gives a particularly interesting perspective on the limitations of most actual interactive music systems.

The classical approach to cognitive science assumes that external behaviour is mediated by internal representations (Fodor 1975) and that cognition is basically the manipulation of these mental representations by sets of rules. It mainly relies on the sense-think-act framework (Pfeifer and Scheier 2001), where future actions are planned according to perceptual information.

Interestingly enough, a parallel can be drawn between classical cognitive science and the development of classical music which also heavily relies on the use of formal structures. It puts the emphasis on internal processes (composition theory) to the detriment of the environment or the body, with a centralized control of the performance (the conductor).

Disembodiment in classical music composition can be seen at several levels. Firstly, by training, the composer is used to composing in his head and translating his mental representations into an abstract musical representation: the score. Secondly, the score is traditionally interpreted live by the orchestra's conductor who 'controls' the main aspects of the musical interpretation, whereas the orchestra musicians themselves are left with a relatively reduced interpretative freedom. Moreover, the role of audience as an active actor of a musical performance is mostly neglected.

An alternative to classical cognitive science is the connectionist approach that tries to build biologically plausible systems using neural networks. Unlike more traditional digital computation models based on serial processing and explicit manipulation of symbols, connectionist networks allow for fast parallel computation. Moreover, it does not rely on explicit rules but on emergent phenomena stemming from the interaction between simple neural units. Another related approach, called embodied cognitive science, put the emphasis on the influence of the environment on internal processes. In some sense it replaced the view of cognition as a representation by the view that cognition is an active process involving an agent acting in the environment. Consequently, the complexity of a generated structure is not the result of the complexity of the underlying system only but partly due to the complexity of its environment (Simon 1981).

Musical counterparts of some of these ideas can be found in American experimental music and most notably in John Cage's work. For instance, the famous 4'33" silent piece transposes the focus of the composition from a strict interpretation of the composer's score to the perception and interpretation of the audience itself. The piece is shaped by the noise in the audience, the acoustics of the performing hall, the reaction of the environment. Cage also made heavy use of probabilities and chance operations to compose some of his pieces. For instance he delegated the 'central control' approach of traditional composers to the aleatory rules of the traditional Chinese I Ching divination system in 'Music of Changes'.

Another interesting aspect of American experimental music is how minimalist music composers managed to create complexity from small initial variations of basic musical material. This can be directly put into relation with the work of Braintenberg on robot/vehicles which appear to have seemingly intelligent behaviours while being governed by extremely simple laws (Braitenberg 1984). A striking example is the use of phase delays in compositions by Steve Reich. In 'Piano Phase', Reich mimics with two pianists the effect of dephasing two tapes playing the same material. Even if the initial pitch material and the phasing process are simple, the combination of both gives rise to the emergence of a complex and interesting musical piece mediated by the listener's perception.

A piece that gives a good illustration of situatedness, distributed processing, and emergence principles is 'In C' by Terry Riley. In this piece, musicians are given a set of pitch sequences composed in advance, but each musician is left in charge of choosing when to start playing and repeating these sequences (Figure 2). The piece is formed by the combination of decisions of each independent musician that makes her decision based on the collective musical output that emerges from all the possible variations.

Following recent evolution of our understanding of cognitive systems, we want to emphasize the crucial role of emergence, distributed processes and situatedness (as opposed to rule-based, serial, central, internal models) in the design of interactive music composition systems.

Figure 2: The score of 'In C' by Terry Riley.

3.2 Human in the loop

In the context of an interaction between a music system and the user, one relevant aspect is personal enjoyment, excitement and well-being as described in the theory of Flow by (Csikszentmihalyi 1991). As a matter of fact, flow and creativity have been found to be related as a result of musical interaction (MacDonald *et al.* 2006; Pachet 2006).

Csikszentmihalyi's theory of Flow is an attempt at understanding and describing the state of Flow (or optimal experience) experienced by creative people. It takes a subjective viewpoint on the problem and describes creativity as a personal feeling of creating something new and interesting in a specific context of production. One interesting aspect of the theory of Flow is that it relates creativity to a certain well-being obtained through an interactive process.

This raises the question of the nature of the human feedback that is injected in a given interactive music system. Indeed, Csikszentmihalyi's theory suggests that the feedback should convey information about the emotional state of the human interactor in order to create an interesting flow-like interaction. This means the design of appropriate interfaces plays a major role in the success of an interactive creative system.

The advent of new sensing technologies has fostered the development of new kinds of interfaces for musical expression. Graphical User Interfaces, tangible interfaces, gesture interfaces have now become omnipresent in the design of live music performance or compositions (Paradiso 2002). For instance, graphical-based software such as IanniX (Coduys and Ferry 2004) or IXI (Magnusson 2005) propose new types of complex multidimensional multimedia scores to the composer. A wealth of gesture-based interfaces have also been devised. A famous example is the project 'The Hands', created by Waisvisz. 'The Hands' is a gestural interface that converts movements of the hands, fingers and arms into sound (Krefeld and Waisvisz 1990). Similarly, 'The Very Nervous System' created by Rokeby transforms dance movements into sonic events (Rokeby 1998). Machover in his large scale 'Brain Opera' project devised a variety of novel interfaces used for the 'Mind Forest' performance (Paradiso 1999). More recently, tangible interfaces, such as the Reactable, which allows a user to interact with digital information through physical manipulation, have become increasingly popular. Most of these interfaces are gesture-based interfaces that require explicit conscious body movements from the user. They do not have access to implicit emotional states of the user.

Although the idea is not new (Knapp and Lusted 1990; Rosenboom 1989), the past few years have witnessed a growing interest from the computer music community in using physiological data such as heart rate, electrodermal activity, electroencephalogram and respiration to generate or transform sound and music. Thanks to the development of more robust and accurate biosignal technologies, it is now possible to derive emotion-related information from physiological data and use it as an input to interactive music systems.

Heart activity measurement has a long tradition in emotion and media research, where it has been shown to be a valid real-time measure for attention and arousal (Lang 1990). Attention evokes short-term (phasic component) deceleration of heart rate, while arousing stimuli accelerate heart rate in the longer term (tonic component). Heart rate change has also been shown to reflect stimuli valence. While the heart rate drops initially after presentation of the stimuli due to attention shift, negative stimuli result in a larger decrease of a longer duration (Bradley and Lang 2000).

Similarly, the study of brainwaves has a rich history, and different brainwave activities have been shown to be correlated with different states. For instance, an increase of energy in the alpha wave frequency typically correlates with states of relaxation (Nunez 2005).

In the literature, we distinguish three main trends in using biosignals. First is the use of physiology to modulate pre-recorded samples, to directly map physiological data to synthesis parameters, or to control higher level musical structures with parameters extracted from the physiology. A popular example of the first category is the Fraunhofer StepMan sensing and music playback device (Bieber and Diener 2005) that adapts the tempo of the music to the speed and rhythm of joggers' steps, calculated from biosensoric data. While this approach appears efficient and successful, it allows control over only one simple musical parameter. The creative possibilities are somewhat limited. In other work by Arslan *et al.* (2006), the emphasis is put on the signal processing chain for analysing the physiological data, which in turn is sonified, using adhoc experimental mappings. Although raw data sonification can lead to engaging artistic results, these approaches do not use higher level interpretation of the data to control musical parameters. Finally, musicians and researchers have used physiological data to modulate the activity of groups of predefined musical cells (Hamilton 2006) containing pitch, metre, rhythm and instrumentation material. This approach allows for interesting and original musical results, but the relation between the emotional information contained in the physiological data and the composer's intention is usually not clearly investigated. Yet, providing emotion-based physiological interfaces is highly relevant for a number of applications including music therapy, diagnosis, interactive gaming, and emotion-aware musical instruments.

Music and its effect on the listener has long been a subject of fascination and scientific exploration, from the Greeks speculating on the acoustic properties of the voice (Kivy 2002) to Musak researchers designing 'soothing' elevator music. It has now become an omnipresent part of our day to day life, whether by choice when played on a personal portable music device, or imposed when diffused in malls during shopping hours for instance. Music is well known for affecting human emotional states, and most people enjoy music because of the emotions it evokes. Yet, the relationship between specific musical parameters and emotional responses is not clear. Curiously, although emotions seem to be a crucial aspect of music listening and performance, the scientific literature on music and emotion is scarce if compared to music cognition or perception (Meyer 1956; Gabrielsson and Lindström 2001; Le Groux *et al.* 2008; Krumhansl 1997; Bradley and Lang 2000; Le Groux and Verschure 2010a). We believe that in order to be complete, the design of a situated music system should take into consideration the emotional aspects of music, especially as the notion of well-being appears to be directly related to flow. Biosignal interfaces in this respect can provide valuable information about the human interactor to the system.

An important decision in the design of a music system is the question of relevant representations. How do changes in technical parameters relate to an actual change at the perceptual level for the listener? Whereas macro-level musical parameters such as pitch, intensity, rhythm and tempo are quite well understood and can be, to a first approximation, modelled and controlled with the MIDI protocol (Anderton 1987), the micro-structure of a sound, its timbre, is not as easy to handle in an intuitive way.

One of the most important shifts in music technology over the last decades was the advent of digital signal processing techniques. Thanks to faster processors, the direct generation of sound waves from compact mathematical representations became reality. Recent years have seen the computer music community focus its efforts in the direction of synthesis of sonic events and the transformation of sound material. Personal computers are now able to synthesize high quality sounds, and sound synthesis software has become largely accessible.

Nevertheless, the use of these tools can be quite intimidating and even counter-intuitive for non-technically oriented users. Building new interesting synthesized sounds, and controlling them, often requires a high level of technical expertise. One of the current challenges of sound and music computing is to find ways to control synthesis in a natural, intuitive, perceptually meaningful manner. Most of the time the relation between a change of synthesis parameter and its effect on the perception of the synthesized sound is not predictable. Due to the high dimensionality of timbre, the automated control of sound synthesis in a generic interactive music system remains a difficult task. The study of the relationship between changes in synthesis parameters and their perceptual counterpart is a crucial question to address for designing 'meaningful' interactive systems.

3.3 'Verum Ipsum Factum': The Synthetic Method

'Verum et factum convertuntur' or 'the true and the made are convertible' is the motto of the synthetic approach proposed by Giambattista Vico (Vico 1725/1862), an early eighteenth-century philosopher. The synthetic approach states that meaning and knowledge is a human construction and that the manipulation of parameters and structure of a man-made synthetic artefact helps to understand the underlying model. For Vico the building process itself is a source of knowledge ('understanding by building', Pfeifer and Scheier 2001; Verschure 1998), as it forces us to think about the role of each element and its interaction with the other parts of the system.

Applying the synthetic approach to engineering (sometimes called 'forward engineering') is not as common as the 'reverse engineering' methodology, but is a good method to avoid the so-called frame of reference problem (Pfeifer and Scheier 2001; Verschure 2002; 1997; 1998). As a matter of fact, when functional analysis (or reverse engineering) is performed, usually all the complexity is assumed to pertain to the cognitive processes, while the role of the environment is underestimated. This is the frame of reference problem. As a result, it has been argued that theories that are produced via analysis are often more complicated than necessary (Braitenberg 1984).

'Analysis is more difficult than invention in the sense in which, generally, induction takes more time to perform than deduction: in induction one has to search for the way, whereas in deduction one follows a straightforward path' (*ibid.*).

When a complex behaviour emerges, the synthetic approach allows the researcher to generate simpler explanations because she knows the proprieties of the components of the system she built. This motivates the choice of a synthetic approach to the study of music perception, cognition and emotion.

4 An approach inspired by situated robotics

The Roboser project is an interesting approach that tackles the problem of music generation using a real-world behaving device (e.g. a robot equipped with sensors) as an input to a MIDI sequencer called Curvasom (Manzolli and Verschure 2005). This way, the musical output somehow illustrates how the robot experiences the world. The robot behaviour is controlled by the Distributed Adaptive Control model (DAC; Verschure *et al.* 2003), a model of classical and operant conditioning, which is implemented using the real-time neuronal simulation environment IQR (Bernardet *et al.* 2002). DAC consists of three layers of control, namely the reactive layer, the adaptive layer and the contextual layer. While the reactive layer is a set of prewired reflex loops, the adaptive layer associates co-occurring stimuli. Finally, the contextual layer provides mechanisms for short- and long-term memory that retain sequences of perceptions/actions that led to a goal state (for instance reaching a light source). Specific neural states such as exploration, collision or light encounter are used to trigger voices or modulate the sequencing parameters (pitch transposition, volume, tempo, velocity).

The aim of Roboser is to integrate sensory data from the environment in real-time and interface this interpreted sensory data combined with the internal states of the control system to Curvasom. The variation in musical performance is provided by the operational states of the system. The more the robot behaves in the environment, the more it learns about this environment, and starts structuring its behaviour. In this way, unique emergent behaviours are generated and mapped to musical parameters. Experiments have shown that the dynamics of a real-world robot exploring the environment induces novelty in the fluctuations of sound control parameters (Manzolli and Verschure 2005).

While the Roboser project paves the way for a new type of interactive music systems based on emergence, parallelism, and the interaction with the environment, there is room for improvement in some of its aspects.

One weakness of the system is that the 'structure generator', i.e. the robot, controlled by DAC (Verschure *et al.* 2003), behaving in the real world, doesn't take into account any musical feedback. In this paradigm, from the robot's perspective, the learning of perception/action sequences depends on the structure of the robot arena only, not on the musical output. The music is driven by a fixed one-way mapping from spatial behaviour to a reduced set of expressive MIDI parameters (volume, tempo, velocity, pitch bend). There is no interaction between the behaviour of the robot and the musical output, as there is no music-related feedback sent to the robot. Hence, the musicality or expressive quality of the result is not taken into account by the system. The human listener is not taken into account in this model either and does not contribute any emotional or musical feedback. The robot, the sequencer Curvasom, and the listener are somewhat connected but do not really interact. Moreover, Curvasom can only generate fixed MIDI sequences. It does not allow for control over the micro-level of sound (sound synthesis), nor does it allow to interactively change the basic musical content as the piece evolves (for each session, the musical sequences to be played are pre-composed and fixed once for all). Curvasom does not support polyphonic voices, which means musical concepts such as chords cannot be used on a single channel. These limitations put some restrictions on the expressive power of the system, as well as on the musical styles that can be produced, and heavily relies on pre-composed material. At a more technical level, Roboser is not multi-plaform, and the music sequencer Curvasom can only be controlled from the neural simulator IQR (Bernardet and Verschure 2010), which is not the ideal control platform in situations where neural modelling is not deemed necessary.

5 SMuSe: The Situated Music Server

On the one hand, the evolution of computer-based music systems has gone from computer-aided composition, which transpose the traditional paradigms of music composition to the digital realm, to complex feedback systems that allow for rich multimodal interactions. On the other hand, the paradigms on which most interactive music systems relied until now are based on outdated views in the light of modern situated cognitive system design. Moreover, the role of human emotional feedback is rarely taken into account in the interaction loop. Even if the development of modern audio signal processing techniques now allow for efficient synthesis and transformation of audio material directly, the perceptual control of the many dimensions of musical timbre remains an open problem. We propose to address these limitations by introducing a novel synthetic interactive composition system called the SMuSe (Situated Music Server) based on the principles of parallelism, emergence, embodiment and emotional feedback.

5.1 Perceptually and cognitively motivated representations of music

Over the last centuries, most composers in the Western classical music tradition have relied on a standard representation of music (score) that specifies musical dimensions such as tempo, metre, notes, rhythm, expressive indications (crescendi, legati, etc.) and instrumentation. Nowadays, powerful computer music algorithms that enable direct manipulation of properties of a sound wave can run on standard laptops, and the use of extended playing modes (for instance the use of subharmonics on a violin—Kimura 1999—or the use of the instrument's body as a percussive instrument) has become common practice. As the amount of information needed to describe subtle music modulations or complex production techniques increases, musical scores get more sophisticated, and sometimes even include direct specific information concerning the production of the sound waveform itself.

This raises the question of the representation of music. What are the most relevant dimensions of music? Here, we take a cognitive psychology approach, and define a set of parameters that are the most perceptually salient, the most meaningful cognitively. Music is a real-world stimulus that is meant to be perceived by a human listener. It involves a complex set of perceptive and cognitive processes that take place in the central nervous system. These processes are partly interdependent, are integrated in time and involve memory as well as emotional systems (Koelsch and Siebel 2005; Peretz and Coltheart 2003; Peretz and Zatorre 2005). The understanding of these processes shed a light on the structures and features that are perceptually and cognitively relevant. Experimental studies have found that musical perception happens at three different time scales; namely the event fusion level, when basic musical events emerge (pitch, intensity, timbre); the melodic and rhythmic grouping, when patterns of those basic events are perceived; and finally the form level, that deals with large scale sections of music (see Snyder 2000, for a review of music and memory processes). This hierarchy of three time scales of music processing form the basis on which we built the SMuSe music processing chain.

5.2 A bio-mimetic architecture

At the low event fusion level, the SMuSe provides a set of synthesis techniques validated by psychoacoustical tests (Le Groux and Verschure 2009b; Le Groux *et al.* 2008) that gives perceptual control over the generation of timbre as well as the use of MIDI information to define basic musical material such as pitch, velocity and duration. Inspired by previous works on musical performance modelling (Friberg *et al.* 2006), SMuSe allows us to modulate the expressiveness of music generation by varying parameters such as phrasing, articulation and performance noise (Le Groux and Verschure 2009b). These nuances are fundamentally of a continuous type unlike pitch or rhythm (Snyder 2000). They cannot be easily remembered by listeners and are typically processed at the level of echoic memory (Raffman 1993).

At the medium melodic and rhythmic grouping level, SMuSe implements various state-of-the-art algorithmic composition tools (e.g. generation of tonal, Brownian and serial series of pitches and rhythms, Markov chains, etc.) The time scale of this mid-level of processing is in the order of 5 seconds for a single grouping, i.e. the time limit of auditory short-term memory.

The form level concerns large groupings of events over a long period of time (longer than the short-term memory). It deals with entire sequences of music and relates to the structure and limits of long-term memory. This longer term structure is accomplished via the interaction with the environment.

SMuSe follows a bio-mimetic architecture that is multi-level and loosely distinguishes sensing (e.g. electrodes attached to the scalp using a cap) from processing (musical mappings and processes) and actions (changes of musical parameters). It has to be emphasized though that we do not believe that these stages are discrete modules. Rather, they will share bi-directional interactions both internal to the architecture as through the environment itself. In this respect it is a further advance from the traditional separation of sensing, processing and response paradigm (Rowe 1993) which was at the core of traditional AI models (Verschure et al. 2003).

SMuSe is implemented as a set of Max/MSP abstractions and C++ externals (Zicarelli 2002) that implement a cognitively plausible system. It relies on a hierarchy of perceptually and musically meaningful agents (Minsky 1988) that can communicate via the OSC protocol (Wright 2005). SMuSe can interact with the environment in many different ways and has been tested with a variety of sensors such as biosignals like heart-rate or electroencephalogram (Le Groux and Verschure 2009a,b; Le Groux et al. 2008), or virtual and mixed-reality sensors like camera, gazers, lasers, and pressure sensitive floors (Bernardet et al. 2009).

The use of the OSC protocol for addressing agents means that the musical agents can be controlled and accessed from anywhere (including over a network if necessary) at any time. This gives great flexibility to the system, and allows for shared collaborative compositions where several clients can access and modulate the music server. In this collaborative composition paradigm, every performer builds on what the others have done. The result is a complex sound structure that keeps evolving as long as different performers contribute changes to its current shape. A parallel could be drawn with stigmergic mechanisms of coordination between social insects like ants (Simon 1981; Bonabeau et al. 1999; Hutchins and Lintern 1995). In ant colonies, the pheromonal trace left by one ant at a given time is used as a means to communicate and stimulate the action of the others. Hence they manage to collectively build complex networks of trails towards food sources.

Similarly, in a collective music paradigm powered by an OSC client/server architecture, one performer leaves a musical trace to the shared composition, which in turn stimulates the other co-performers to react and build on top of it.

5.3 Human feedback

SMuSe has been used in various sensing environments, where sensory data is used to modulate SMuSe's musical output. Yet, in order to reinject specific music-based feedback into SMuSe, the only solutions are whether to build a sophisticated music analysis agent, or to somehow measure a human listener's response to the musical output. Most people acknowledge they listen to music because of its emotional content. Hence the choice of musical emotion as a feedback signal seems a natural choice.

In the context of research on music and emotion, one option is to exploit the vast amount of research that has been conducted to investigate the relationship between specific musical parameters and emotional responses (Gabrielson and Juslin 1996; Juslin *et al.* 2001). This gives a set of reactive, explicit mappings (*cf.* Table 1).

Musical Parameter	Level	Semantics of Musical Expression
Tempo	slow	Sadness, Calmness, Dignity, Boredom
	fast	Happiness, Activity, Surprise, Anger
Mode	Minor	Sadness, Dreamy, Anger
	Major	Happiness, Grace, Serenity
Volume	Loud	Joy, Intensity, Power, Anger
	Soft	Sadness, Tenderness, Solemnity, Fear
Register	High	Happiness, Grace, Excitement, Anger, Activity
	Low	Sadness, Dignity, Solemnity, Boredom
Tonality	Tonal	Joyful, Dull
	Atonal	Angry
Rhythm	Regular	Happiness, Dignity, Peace
	Irregular	Amusement, Uneasiness, Anger

Table 1: Review of the emotional impact of several musical features (adapted from Juslin *et al.* 2001).

Another possibility is to learn the mappings online as the interaction takes place. This is possible in SMuSe thanks to a specialized reinforcement learning agent (Sutton and Barto 1998). Reinforcement learning is particularly suited for an explorative and adaptive approach to mapping, as it tries to find a sequence of parameter changes that optimizes a reward function. For instance, this principle was tested using musical tension levels as the reward function (Le Groux and Verschure 2010b). Interestingly enough, the biological validity of reinforcement learning is supported by numerous studies in psychology and neuroscience that found various examples of reinforcement learning in animal behaviour like the foraging behaviour of bees (Montague *et al.* 1995), or the dopamine system in primate brains (Schultz *et al.* 1997).

6 Conclusion

SMuSe illustrates a novel situated approach to music composition systems. It takes advantage of its interaction with the environment to go beyond the classic sense-think-act paradigm (Rowe 1993). It is built on a cognitively plausible architecture that takes into account the different time frames of music processing, and uses an agent framework to model a society of simple distributed musical processes. It combines standard MIDI representation with perceptually-grounded sound synthesis techniques and is based on modern data-flow audio programming practices (Puckette 1996). SMuSe is designed to work with a variety of sensors, most notably physiological. This allows it to re-inject feedback information to the system concerning the current emotional state of the human listener/interactor. SMuSe includes a set of 'pre-wired' emotional mappings from emotions to musical parameters grounded in the literature on music and emotion, as well as a reinforcement learning agent that allows for adaptive mappings. The system design and functionalities have been constantly tested and improved, to adapt to different real-world contexts and has been used during several artistic performances. In the continuation of previous work on Roboser and Ada, a human accessible space that was visited by over 500,000 people (Eng *et al.* 2003), we have further explored the purposive construction of interactive installations and performances. To name but a few, during the VRoboser installation (Le Groux *et al.* 2007a), the sensory inputs (motion, colour, distance, etc.) of a 3D virtual khepera robot living in a game-like environment modulated musical parameters in real time, thus creating a never-ending musical soundscape in the spirit of Brian Eno's 'Music for Airports'. The robot was controlled via a joystick by a participant whose decisions were influenced by both the robot's spatial behaviour and the resulting musical output. This provided an essential human and music feedback loop originally missing in the Roboser paradigm. In another context the SMuSe generated automatic soundscapes and music which reacted to and influenced the spatial behaviour of humans and avatars in a mixed-reality space called XIM, the eXperience Induction Machine (Bernardet *et al.* 2010; Le Groux *et al.* 2007b), thus emphasizing the role of the environment and interaction on the musical composition. Based on similar premises, re(PER)curso, an interactive mixed-reality performance involving dance, percussion, interactive music and video, was presented at the ArtFutura Festival 07 and Museum of Modern Art in Barcelona in the same year. Re(PER)curso was performed in an augmented mixed-reality environment, where the physical and the virtual were not overlapping but distinct and continuous. The border between the two environments were the projection screen that acted like a dynamic all-seeing bi-directional eye. The performance was composed by several interlaced layers of artistic and technological activities — e.g. the music had three components: a predefined soundscape, the percussionist who performed from a score and the interactive composition system; the physical actors, the percussionist and the dancer were tracked by a video-based active tracking system that in turn controlled an array of moving lights that illuminated the scene. The spatial information from the stage obtained by the tracking system was also projected onto the virtual world where it modulated the avatar's behaviour allowing it to adjust body position, posture and gaze to the physical world. Re(PER)curso was operated as an autonomous interactive installation that was augmented by two human performers. In 2009, the 'Brain Orchestra' (Le Groux *et al.* 2010), a multimodal performance using brain-computer interfaces, explored the creative potential of a collection of brains directly interfaced to the world. During the performance four 'brain musicians' were controlling a string quartet generated by the SMuSe using their brain activity alone. The orchestra was conducted by an 'emotional conductor', whose emotional reactions were recorded using biosignal interfaces and fed back to the system. The 'Brain Orchestra' was premiered in Prague for the FET 09 meeting. Central to all these examples of externalist aesthetics has been our new paradigm of situated interactive music composition implemented in the SMuSe. We now are seeking to generalize it towards synthetic multimodal narrative generation. It provides a well-grounded approach towards the development of advanced synthetic aesthetic systems and a further understanding of the fundamental psychological processes on which it relies.

References

Ames, C. (1987), 'Automated Composition in Retrospect: 1956–1986', *Leonardo*, **20** (2): 169–85.

Ames, C. (1989), 'The Markov Process as Compositional Model: A Survey and Tutorial', *Leonardo*, **22** (2): 175–87.

Anderton, C. (1987), 'The MIDI Protocol', *5th International Conference: Music and Digital Technology*.

Arslan, B., A. Brouse, J. Castet, P. Lehembre, C. Simon, J.J. Filatriau, and Q. Noirhomme (2006), 'A Real Time Music Synthesis Environment Driven with Biological Signals', *Proceedings of IEEE International Conference on Acoustics, Speech and Signal Processing*, **2**: II–II.

Assayag, G. and S. Dubnov (2002), 'Universal Prediction Applied to Stylistic Music Generation' in G. Assayag, H.G. Feichtinger and J.F. Rodrigues, Eds., *Mathematics and Music: A Diderot Mathematical Forum*, Berlin, Springer Verlag: 147–60.

Assayag, G., G. Bloch, M. Chemillier, A. Cont and S. Dubnov (2006), 'OMax Brothers: A Dynamic Yopology of Agents for Improvization Learning', *Proceedings of the 1st ACM Workshop on Audio and Music Computing Multimedia*: 132.

Bernardet, U. and P.F.M.J. Verschure (2010), 'IQR: A Tool for the Construction of Multi-Level Simulations of Brain and Behavior', *Neuroinformatics*, **8**: 1–22.

Bernardet, U., S. Bermúdez i Badia, A. Duff, M. Inderbitzin, S. Groux, J. Manzolli, Z. Mathews, A. Mura, A. Valijamae and P.F.M.J. Verschure (2010), 'The Experience Induction Machine: A New Paradigm for Mixed-Reality Interaction Design and Psychological Experimentation' in E. Dubois, L. Nigay and P. Gray, Eds., *The Engineering of Mixed Reality Systems*, Berlin, Springer: 357–79.

Bernardet, U., M. Blanchard and P.F.M.J. Verschure (2002), 'IQR: A Distributed System for Real-Time Real-World Neuronal Simulation', *Neurocomputing*, **44–46**: 1043–8.

Bernardet, U., S. Bermúdez i Badia, A. Duff, M. Inderbitzin, S. Le Groux, J. Manzolli, Z. Mathews, A. Mura, A. Valjamae and P.F.M.J. Verschure (2009), *The eXperience Induction Machine: A New Paradigm for Mixed Reality Interaction Design and Psychological Experimentation*, Berlin, Springer.

Bieber, G. and H. Diener (2005), *StepMan–A New Kind of Music Interaction*, Mahwah (NJ), Lawrence Erlbaum Associates.

Bonabeau, E., M. Dorigo and G. Theraulaz (1999), *Swarm Intelligence: From Natural to Artificial Systems*, New York, Oxford University Press.

Bongers, B. (2000), 'Physical Interfaces in the Electronic Arts: Interaction Theory and Interfacing Techniques for Real-Time Performance', *Trends in Gestural Control of Music*: 41–70.

Bradley, M.M. and P.J. Lang (2000), 'Affective Reactions to Acoustic Stimuli', *Psychophysiology*, **37** (2): 204–15.

Braitenberg, V. (1984), *Vehicles: Explorations in Synthetic Psychology*, Cambridge (MA), MIT Press.

Coduys, T. and G. Ferry (2004), 'Iannix Aesthetical/Symbolic Visualisations for Hypermedia Composition', *Proceedings International Conference Sound and Music Computing*.

Cope, C. (1996), *Experiments in Musical Intelligence*, Middleton (WI), A-R Editions.

Csikszentmihalyi, M. (1991), *Flow: The Psychology of Optimal Experience*, London, Harper Perennial.

Eng, K., A. Babler, U. Bernardet, M. Blanchard, M. Costa, T. Delbruck, R.J. Douglas, K. Hepp, D. Klein, J. Manzolli, M. Mintz, F. Roth, U. Rutishauser, K. Wassermann, A.M. Whatley, A. Wittmann, R. Wyss and P.F.M.J. Verschure (2003), 'Ada—Intelligent Space: An Artificial Creature for the Swissexpo.02.', *IEEE International Conference on Robotics and Automation, Proceedings. ICRA '03*, **3**: 4154–9.

Fodor, J.A. (1975), *The Language of Thought*, Cambridge (MA), Harvard University Press.

Friberg, A., R. Bresin and J. Sundberg (2006), 'Overview of the kth Rule System for Musical Performance', *Advances in Cognitive Psychology*, **2** (2–3):145–61.

Gabrielson, A. and P.N. Juslin (1996), 'Emotional Expression in Music Performance: Between the Performer's Intention and the Listener's Experience', *Psychology of Music*, **24**: 68–91.

Gabrielsson, A. and E. Lindström (2001), 'Music and Emotion—Theory and Research' in P. Juslin and J. Sloboda, Eds., *The Influence of Musical Structure on Emotional Expression*, New York, Oxford University Press: 223–48.

Hamilton, R. (2006), 'Bioinformatic Feedback: Performer Bio-Data as a Driver for Real-Time Composition', *Nime '06: Proceedings of the 6th International Conference on New interfaces for Musical Expression*.

Hutchins, E. and G. Lintern (1995), *Cognition in the Wild*, Cambridge (MA), MIT Press.

Jewell, E.J., F.R. Abate and E. McKean (2001), *The New Oxford American Dictionary*, New York, Oxford University Press.
Juslin, P.N., J.A. Sloboda and Anonymous (2001), *Music and Emotion: Theory and Research*, New York, Oxford University Press.
Kimura, K. (1999), 'How to Produce Subharmonics on the Violin', *Journal of New Music Research*, **28** (2):178–84.
Kivy, P. (2002), *Introduction to a Philosophy of Music*, New York, Oxford University Press.
Knapp, B.R. and H.S. Lusted (1990), 'A Bioelectric Controller for Computer Music Applications', *Computer Music Journal*, **14** (1): 42–7.
Koelsch, S. and W.A. Siebel (2005), 'Towards a Neural Basis of Music Perception', *Trends in Cognitive Sciences*, **9** (12): 578–84.
Krefeld, V. and M. Waisvisz (1990), 'The Hand in the Web: An Interview with Michel Waisvisz', *Computer Music Journal*, **14** (2): 28–33.
Krumhansl, C.L. (1997), 'An Exploratory Study of Musical Emotions and Psychophysiology', *Canadian Journal of Experimental Psychology*, **51** (4): 336–53.
Lang, A. (1990), 'Involuntary Attention and Physiological Arousal Evoked by Structural Features and Emotional Content in TV Commercials', *Communication Research*, **17**: 275–99.
Le Groux, S., J. Manzolli and P.F.M.J. Verschure (2007a), 'VR-RoBoser: Real-Time Adaptive Sonification of Virtual Environments Based on Avatar Behaviour' in *NIME '07: Proceedings of the 7th International Conference on New Interfaces for Musical Expression*, New York: ACM Press: 371–4.
Le Groux, S., J. Manzolli and P.F.M.J. Verschure (2007b), 'Interactive Sonification of the Spatial Behavior of Human and Synthetic Characters in a Mixed-Reality Environment', *Proceedings of the 10th Annual International Workshop on Presence*, Barcelona, Spain.
Le Groux, S., J. Manzolli, M. Sanchez, A. Luvizotto, A. Mura, A. Valjamae, C. Guger, R. Prueckl, U. Bernardet and P.F.M.J. Verschure (2010), 'Disembodied and Collaborative Musical Interaction in the Multimodal Brain Orchestra', *NIME '10: Proceedings of the International Conference on New Interfaces for Musical Expression*, Sydney, Australia.
Le Groux, S., A. Valjamae, J. Manzolli and P.F.M.J. Verschure (2008), 'Implicit Physiological Interaction for the Generation of Affective Music', *Proceedings of the International Computer Music Conference*, Belfast.
Le Groux, S. and P.F.M.J. Verschure (2009a), 'Neuromuse: Training Your Brain Through Musical Interaction', *Proceedings of the International Conference on Auditory Display*.
Le Groux, S. and P.F.M.J. Verschure (2009b), 'Situated Interactive Music System: Connecting Mind and Body Through Musical Interaction', *Proceedings of the International Computer Music Conference*, Montreal, Canada.
Le Groux, S. and P.F.M.J. Verschure (2010a), 'Emotional Responses to the Perceptual Dimensions of Timbre: A Pilot Study Using Physically Inspired Sound Synthesis', *Proceedings of the 7th International Symposium on Computer Music Modeling*, Malaga, Spain.
Le Groux, S. and P.F.M.J. Verschure (2010b), 'Adaptive Music Generation by Reinforcement Learning of Musical Tension', *7th Sound and Music Computing Conference*, Barcelona, Spain.
Lerdahl, F. and R. Jackendoff (1996), *A Generative Theory of Tonal Music*, Cambridge (MA), MIT Press.
MacDonald, R., C. Byrne and L. Carlton (2006), 'Creativity and Flow in Musical Composition: An Empirical Investigation', *Psychology of Music*, **34** (3): 292–306.
Machover, T. and J. Chung (1989), 'Hyperinstruments: Musically Intelligent and Interactive Performance and Creativity Systems', *Proceedings of the 1989 International Computer Music Conference*.
Magnusson, T. (2005), 'IXI Software: The Interface as Instrument', *Proceedings of the 2005 Conference on New Interfaces for Musical Expression*, Singapore: 212–5.
Manzolli, J. and P.F.M.J. Verschure (2005), 'Roboser: A Real-World Composition System', *Journal of Computer Music*, **29** (3): 55–74.
Meyer, L.B. (1956), *Emotion and Meaning in Music*, Chicago, University of Chicago Press.
Minsky, M. (1988), *The Society of Mind*, New York, Simon and Schuster.

Montague, P.R., P. Dayan, C. Person and T.J. Sejnowski (1995), 'Bee Foraging in Uncertain Environments Using Predictive Hebbian Learning', *Nature*, **377** (6551): 725–8.

Narmour, E. (1990), *The Analysis and Cognition of Basic Melodic Structures: The Implication-Realization Model*, Chicago, University of Chicago Press.

Nunez, P.L. (2005), *Electric Fields of the Brain: The Neurophysics of EEG*, New York, Oxford University Press.

Pachet, E. (2006), 'Enhancing Individual Creativity with Interactive Musical Reflective Systems' in I. Deliège and G. Wiggins, Eds., *Musical Creativity: Multidisciplinary Research in Theory and Practice*, London, Psychology Press.

Paradiso, J.A. (1999), 'The Brain Opera Technology: New Instruments and Gestural Sensors for Musical Interaction and Performance', *Journal of New Music Research*, **28** (2): 130–49.

Paradiso, J.A. (2002), 'Electronic Music: New Ways to Play', *Spectrum IEEE*, **34** (12): 18–30.

Peretz, I. and M. Coltheart (2003), 'Modularity of Music Processing', *Nature Neuroscience*, **6** (7): 688–91.

Peretz, I. and R.J. Zatorre (2005), 'Brain Organization for Music Processing', *Psychology*, **56**: 89–114.

Pfeifer, R. and C. Scheier (2001), *Understanding Intelligence*, Cambridge (MA), MIT Press.

Puckette, M. (1996), 'Pure Data: Another Integrated Computer Music Environment', *Proceedings of the 2nd Intercollege Computer Music Concerts*: 37–41.

Raffman, D. (1993), *Language, Music, and Mind*, Cambridge (MA), MIT/Bradford Books.

Rokeby, S., Ed. (1998), *Construction of Experience. Digital Illusion: Entertaining the Future with High Technology*, New York, ACM.

Rosenboom, D. (1989), 'Biofeedback and the Arts: Results of Early Experiments', *Computer Music Journal*, **13**: 86–8.

Rowe, R. (1993), *Interactive Music Systems: Machine Listening and Composing*, Cambridge (MA), MIT Press.

Sacks, O. (2008), *Musicophilia: Tales of Music and the Brain*, New York, Alfred A. Knopf.

Schultz, W., P. Dayan and P.R. Montague (1997), 'A Neural Substrate of Prediction and Reward', *Science*, **275** (5306): 1593.

Simon, H.A. (1981), *The Science of the Artificial*, Cambridge (MA), MIT Press.

Snyder, B. (2000), *Music and Memory: An Introduction*, Cambridge (MA), MIT Press.

Sutton, R.S. and A.G. Barto (1998), *Reinforcement Learning: An Introduction (Adaptive Computation and Machine Learning)*, Cambridge (MA), MIT Press.

Varela, F., E. Thompson and E. Rosch (1991), *The Embodied Mind*, Cambridge (MA), MIT Press.

Verschure, P.F.M.J. (1997), 'Connectionist Explanation: Taking Positions in the Mind-Brain Dilemma' in G. Dorffner, Ed., *Neural Networks and a New Artificial Intelligence*, London, Thompson: 133–88.

Verschure, P.F.M.J. (1998), 'Synthetic Epistemology: The Acquisition, Retention, and Expression of Knowledge in Natural and Synthetic Systems', *The 1998 IEEE International Conference on Fuzzy Systems Proceedings*, **1**: 147–52.

Verschure, P.F.M.J. (2002), 'Formal Minds and Biological Brains: AI and Edelman's Extended Theory of Neuronal Group Selection', *IEEE Expert*, **8** (5): 66–75.

Verschure, P.F.M.J., T. Voegtlin and R.J. Douglas (2003), 'Environmentally Mediated Synergy Between Perception and Behaviour in Mobile Robots', *Nature*, **425** (6958): 620–4.

Vico, G. (1725/1862), *Principi di Scienza Nuova*, Milan, Librajo-Editore Fortunato Perelli.

Winkler, T. (2001), *Composing Interactive Music: Techniques and Ideas Using Max*, Cambridge (MA), MIT Press.

Wright, M. (2005), 'Open Sound Control: An Enabling Technology for Musical Networking', *Organized Sound*, **10** (3): 193–200.

Zicarelli, D. (2002), 'How I Learned to Love a Program that Does Nothing', *Journal of Computer Music*, **26**: 44–51.

Stéphane Dumas

Creation as Secretion
An Externalist Model in Aesthetics[1]

Artists are now using biotechnologies. How can biological models enrich art and aesthetics? A model relating to cutaneous secretion is proposed in this chapter. Based on some of the characteristics of live skin, it establishes a relationship between language and secretion. It avoids reducing aesthetic experiences to no more than neural processes taking place in the central nervous system. It stands apart from models relying on projection, in particular symbolic projection, which are widespread in cognitive research, among other scientific domains. I use this model first to analyse a robotic sculpture dealing with skin, Alexitimia, by Paula Gaetano Adi, whose approach emphasizes the relationship between brain, body and environment. Skin is a major interface in this relation, being linked both to the body — especially the nervous system — and to the world. After analysing a second artwork — a robotic artist (MEART) — I propose a critique of the notion of representation, establishing a link between different theories of embodiment and the aesthetic model of creation as secretion.

According to Antonin Artaud, human destiny is a kind of ritual. In order to carry out this ritual, 'MAN' must be alone, with no parents, no genealogy, and no transcendent guarantor or tutelage. 'MAN' is an innate quality, endowing a person with both an atavistic and a personal consciousness. This ancestral and individual consciousness is embodied in the skeleton and skin, rather than in the brain:

> [with Peyote] MAN is alone, desperately scraping out the music of his own skeleton, without father, mother, family, love, god, or society.
> And no living being to accompany him. And the skeleton is not of bone but of skin, like a skin that walks. And one walks from the equinox to the solstice, buckling on one's own humanity. (Artaud 1976a, p. 37–8)

---

1 This chapter is partly derived from 'Création et sécrétion', a text published in 'Les arts dans le cadre actuel de la théorie darwinienne de l'évolution', IMÉRA, Marseilles, 2009, http://www.imera.fr/index.php/fr/flux-rss/110.html

This footnote was added by Antonin Artaud to his text 'The Peyote Rite in Tarahumara Country', written during his stay at a psychiatric hospital in Rodez, more than ten years after the author had been initiated in the Peyote ritual by a Mexican Indian tribe.[2] In these lines, Artaud uses a graphic image for human destiny, glimpsed during his initiatory experience. This ritual is carried out by a skin, 'like a skin that walks... buckling on one's own humanity'. Significantly, the writer who coined this skin metaphor for the human destiny also created the concept of a 'body without organs' (Artaud 1976b).[3] The skin is the largest bodily organ. Using this metonymy, one organ stands for the entire human person expressing his or her humanity. However, with the organless body, Artaud is denying the notion of the organism rather than the organs themselves. His metaphysical anatomy, or ritual autopsy, brings into play a skin that is not an organ, part of an organism, but a flayed skin that lives, walks and acts independently of its parent organism. This body-organ foreshadows a body without organs that has been 'delivered... from all its automatic reactions' (Artaud 1976b), an embodied consciousness.

In this paper, the walking skin plays the role of initiator on a journey from the bodily envelope towards what I call 'creative skins',[4] a research model applied to art, based on the physiological properties of the skin, one of which is developed here: secretion. I am suggesting that artistic creation functions like a secretion, but not so as to prevent art from producing worlds using symbolic projection. This switch from nature to culture questions the very distinction between these two categories. At a time when the major trends, notably in cognitive research, involve working on more distant or abstract models relying on symbolic projection, an embodied model based on secretion seems of the utmost importance, especially when applied to art.

1 Skin and language:
An extension beyond the envelope

Cutaneous secretions enable the body to release substances coming from inside the organism or produced by the skin itself. Sweat glands exude in two different ways. Some expel sweat via a duct opening up through a pore whilst others release a thicker substance partly comprising actual glandular material. Transcutaneous secretions contribute to internal bodily regulation (such as temperature, hygrometry or toxins), but they also enable an organism to send signals out to its environment. Such mes-

---

2 This footnote was added in 1947, when Artaud became critical of the mystical tone of the original text written in 1943, at the beginning of his stay at a psychiatric hospital in Rodez. Steven Connor has put this quotation as an epigraph to his *Book of Skin* (2004).
3 Gilles Deleuze and Félix Guattari made an extensive use of this notion (Deleuze and Guattari 1987/2004).
4 *Creative Skins* [*Les peaux créatrices*] is the title of a book to be published in French.

sages can be either defensive or sexual—geared towards attracting other individuals. The production of melanin by the epidermic cells plays a major role in skin coloration and in the perception of physical appearance. In some animal species, it can radically alter the aspect of the skin. Under stressful conditions for instance, the pigment-producing chromatophores can also release a toxic substance through a secretion process. Olfactory traces emitted through the skin are a major tool used by animals staking out their territory, whilst visual signs conveyed by the epidermic surface are fundamental for intersubjective communication. Last but not least, the taste of skin can play a key role in a love relationship.

Are language and physiological secretions radically different processes? Language is a projection whereby one's physical borders are crossed through the sending and receiving of messages. The more a language is linked to a culture, the more richly charged it becomes with symbolism, an encoding system shared by a community and allowing references of varying complexity to things not immediately present. Various mediums conveying language correspond to the different senses, from touch to sight, hearing, smell and taste. Some communication modes work at close quarters and even through direct contact, while others act through long-range projection mostly based on symbolic systems.

In Greek mythology, Apollo is the god of the Logos and mouthpiece of Zeus. This corresponds to his oracular function. He is known as the god 'who strikes from afar' (Homer 1924, I, p. 14).[5] His two main attributes are the bow and the cithara. The latter is no less efficient than the former when it comes to broadcasting his poetic words. This paradigm of remote symbolic projection lays the foundation of language in line with a projective model, which is very common nowadays. Unlike this projective model, the secretion model conveys the dimension of embodiment, which is present in all languages, especially artistic ones. This difference is played out in Greek mythology in the musical competition between Apollo and Marsyas (see Dumas 2008). At the end of the story, the god flays his rival alive. The satyr's animal skin is exhibited after the musician's death. It reacts intelligently when music is played nearby, quivering when Marsyas's music is performed while remaining deaf to Apollo's (Aelian 1997, XIII, 21). This is a major reference for what I call 'creative skins'.

Language oversteps physical boundaries and plays around with the principle of delineation. It can work either close up or at a distance. This difference mostly concerns the technical tools of mediation. However, what differentiates the projection and the secretion models more radically

---

5 *Hekebolos*, 'who strikes from afar', is also translated as 'the far shooter', 'the far darter' or 'the free shooter, according to will'. Apollo's bow is the instrument allowing him to project far away his presence and his father's will, just as does his cithara: 'The lyre and the curved bow shall ever be dear to me, and I will declare to men the unfailing will of Zeus' (Homer 1914, Hymn 131).

is the role played in them by the symbolic and analog modes. The symbolic mode is especially effective for distant projection. Remote transmission has always sought out more powerful encoding systems, inventing the digital language used by computers (see Claude Shannon's fundamental contribution to the theory of information). But analog mode works best with proximity and contact, more or less like a physical imprint. In our cultures the two modes are intermingled.

The secretion and projection models are tools that can help us analyse the effectiveness and the role of art in relation to biology, but they should not be used to explain art by applying some kind of universal biological rule in order to demonstrate the aesthetic effectiveness of particular pieces. Why are some artworks possessed of such aesthetic effectiveness that they can touch audiences beyond the boundaries of time and cultural differences? This question has often been debated in aesthetics. Some writings on neuroaesthetics tend to establish a link between this aesthetic effectiveness and neurobiological laws. This new form of Gestalt relies on some very interesting discoveries, such as mirror neurons, which may play a role in empathy, the ability to feel what someone else is feeling (Rizzolatti and Sinigaglia 2008). However, this approach is often narrowed down to present 'good art' — aesthetically effective art — as following these pre-existing biological rules (Ramachandran 2005; 2007). There are plenty of scientific experiments to prove beyond doubt that an artwork is effective and that such art is 'good art' — but this is precisely what aesthetics is not about!

In my view, aesthetic effectiveness does not come primarily from an artwork's ability to function according to pre-existing rules, but from its capacity to secrete situations whereby the public recreates an aesthetic experience through the artwork's dual nature as both object and process. Artistic creation is not just a projection on a pictorial plane (whatever the material support) of a set of elements that the artist has selected from an available stock in order to embody an idea and an emotion that have developed in his or her mind. Similarly, the reception of art is more than just the projection of a public that perceives the artist's intention through the forms in which it is embodied, triggering reactions that combine emotion and rationality (Changeux 1997, especially 'The cultural imprint'; 2002; 2008; Zeki 2000). The creation of art may equally well be a secretion process producing effects just as geological concretion does, where the resulting forms do not stabilize because they match some pre-existing scheme, pertaining to a sort of cosmic harmony conveyed through genetic transmission, but, on the contrary, do so by polluting any *a priori* form, by contaminating the purity of any line or surface, and by creating unlikely worlds. The reception of art would then rely on contamination as much as

Creation as Secretion

on empathy, whilst the intersubjective communication it affords has to do with tactile contact, porosity and skin exchange.

This approach to aesthetics, which links creation and secretion, and envisions the aesthetic experience as a sort of skin exchange, is closely linked to certain theories of embodiment. After analysing two artworks to ground my arguments, I will introduce some key notions relating to externalism — namely affordances (Gibson 1977), enaction (Varela *et al.* 1991; Noë 2004), extended mind (Clark 2008), also 'brain-body-world nexus' and process (Rockwell 2005, p. 71; Manzotti 2006).

2 Extension through direct touch: Secretion and language

Relying on secretion, the work of art I am now about to analyse shows how both models — the projection and the secretion one — can work together. 'Alexitymia' designates an inability to communicate one's emotions. People suffering from this disability cannot identify their own feelings, nor can they communicate, share, compare or evaluate them. During an emotional conflict, they will not be able to externalize their personal emotions through verbal language, and psychic tensions will appear through symptoms often situated on the skin surface. Erythema and diaphoresis (excessive perspiration) frequently have psychic causes, even in non-pathological situations.

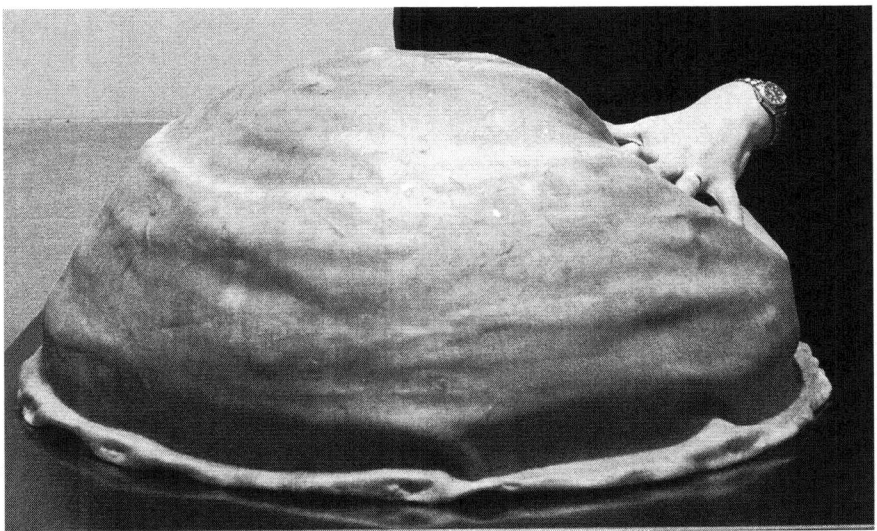

Figure 1: Paula Gaetano Adi, Alexitimia, 2007.

Artist Paula Gaetano Adi has created an artwork titled Alexitimia, which is both an organic sculpture and an autonomous robotic agent.[6] As a sculpture, it is a hemispherical, irregular, flexible shape, made of a skin-like material laid over a rectangular metal base. The piece stands on the floor and is about a metre high. The flexible element looks like an outsize organ that has been placed in a sterilized dish after an autopsy. If you bend down to get a closer look at it, you are tempted to touch it. Its breast-shaped bulge seems to be crying out to be stroked and its flexible surface gently yields to pressure before reverting to its initial volume. Soon something incongruous happens, with small beads of water forming on an increasingly glistening surface. Continued palpation causes this to spread and the casing becomes damp, while the oozing decreases and disappears when the tactile exploration stops.

This occurs owing to the sculpture's robotic dimension. On being touched, pressure sensors embedded in the latex skin of the flexible dome transmit the stimulus to a microcontroller, which activates the excretion process. Pumps inject water into very thin ducts to orifices dotted across the surface of the skin and the robot simulates the physiological process of perspiration. However, its brain—the microcontroller—has not been programmed with sophisticated algorithms aiming at scientific experimentation, such as testing the capacities of an artificial intelligence system in a complex interaction involving unexpected situations encompassing the public's behaviour, the stimuli transmitted by the sensors and the secretory response. The most complex reactions relating to this work are to be observed among the audience, making this piece more of a sculpture than a robot. Its tactile dimension and ability to sweat create a very intimate relationship with the public, who can tell that the sculpture is reacting to someone's touch by sweating.

The fact that an artefact—an artwork at that—can sweat really calls into question our view of a manmade object. Its minimalist and sculptural dimension sets the scene for its real language to emerge somewhat mutedly and barely noticeably, using the language of secretion. Alexitimia's bulging surface is the interface where two different movements meet: perspiration oozes out of its opaque mass, while cultural references are projected by the public looking at the artwork. The two meet through touch. The effectiveness of its language and the uniqueness of this aesthetic experience rely on the subtle and incongruous eruption of a silent secretion.

Alexitimia is a system involving an audience, a sculpture and a robotic process, where the sculpture behaves like an externalized skin. The public's interaction with this living flayed skin suggests a form of reciprocity: I touch and am touched simultaneously. This reflexivity relies on the two

---

[6] Paula Gaetano Adi, *Alexitimia*, 2007; latex, electronic devices, water and steel; ab. 1x1x1 m.

senses of touch and sight. The first contact with the sculpture is through sight; then comes palpation, to reveal what the eye could not see inside this closed shape: its sensitivity. This ability to feel and react to palpation is translated by the robot through a secretion that speaks at once to our touch and sight. Our eyes perceive a symptom—perspiration—denoting the feeling the robot has of being touched by us. Naturally, we know we are dealing with just an uncomplicated robot with mechanical, rather than affective reactions. But we are part of a nexus that brings us in close contact with this object, which becomes an other, reacting, and visibly so, through its silent language. In our everyday lives, we invest considerable affect in all kinds of objects, which become extensions of ourselves. Alexitimia is a work about caring, externalization and reflexivity.

Few artists have worked with perspiration. In 1998, Anne Hamilton exhibited her Weeping Wall at the Museum of Contemporary Art in Lyons. This piece questions our relation to a familiar element of our environment—a gently sweating or weeping wall. In 2007, Yann Marussich created a performance entitled Blue Remix, during which the artist lies completely motionless for one hour in a glass box.[7] Blue sweat gradually oozes out of his skin, covering his sculptural white body with dripping stains. The only apparent movement in this performance is perspiration, made visible through its pictorial dimension. In these examples, language and secretion have much in common. We have a great deal to gain from being aware of the silent and bodily dimensions of language that help us across the border of our skin.

3 Extension through distant touch: Secretion (presentation) or projection (representation)?

We now come to another artwork which is also a robot, but equipped with a much more active body and brain. It has no skin as such, since its envelope has been replaced by representational devices. MEART is a collaborative project hosted by SymbioticA.[8] It is a semi-living artist producing drawings, a biobot comprising an organic brain electronically connected to a body composed of mechanical arms and an eye, which is a camera. The brain is a culture of rat neurons produced in Dr. Steve Potter's lab in Atlanta (Georgia). The arms and eye can be anywhere in the world, provided there is an Internet connection to link them to the brain.

---

7   See Yann Marussich's website: http://www.yannmarussich.ch
8   *MEART – The Semi Living Artist*, by Guy Ben-Ary and Philip Gamblen, hosted by SymbioticA, University of Western Australia, Perth, in collaboration with Dr. Steve Potter's Lab at Georgia Institute of Technology. First exhibited under the title *Fish and Chips* in 2001 at Ars Electronica, Linz. http://www.fishandchips.uwa.edu.au

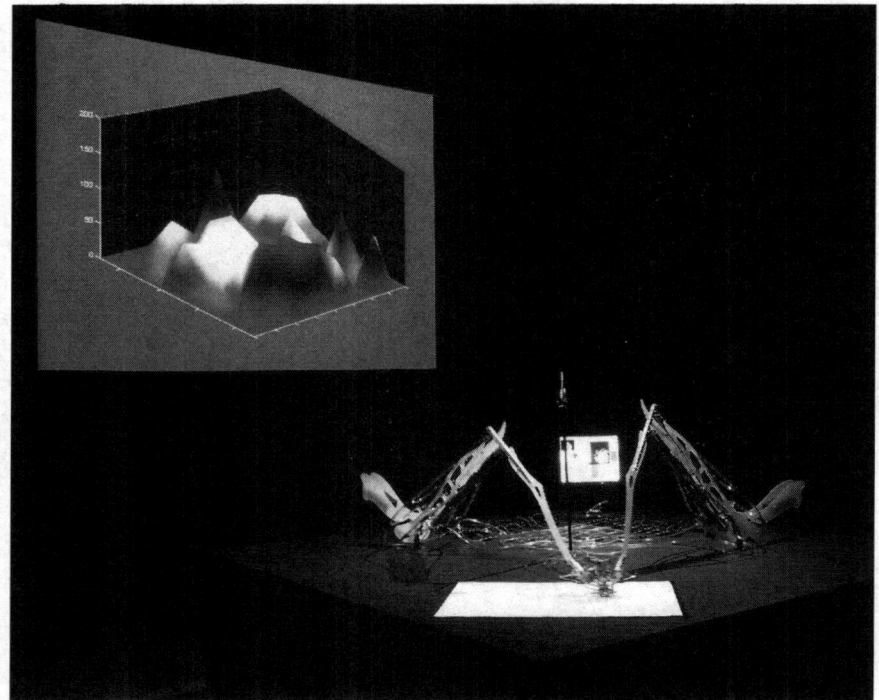

Figure 2: MEART, the robotic arms and a diagram analysing signals from the culture.

First, a digital photograph is taken of an audience member at the exhibition venue, and then communication between the body and the neurons begins. The encoded portrait is transmitted in the form of electric signals to the neuron culture via the Net. The electric neuron reactions to this stimulus are encoded in messages transmitted to the robotic arms, which start moving and drawing. A camera records the progress of the drawing. The image is compared to the initial portrait and a number of error values are transmitted to the neuron culture, which again reacts to this stimulus, and so on. Multiple loops follow each other, and so the drawing progresses until either the cell culture dies or the exhibition closes.

The rat neuron culture is developed on a flat electronic chip called a Multi Electrode Array, with sixty electrodes enabling interaction between the neurons and the robotic body in a kind of sensorimotor relationship. They also enable mapping of the firing neurons. The electronic messages between the neurons and the body transit via computer programs. Neural stimulation is determined comparing the actual drawing with the target image of a person's photograph. The grayscale percentages for corresponding pixels on the two images are continuously compared, in this case subtracted to produce a matrix of sixty error values, which determine the stimulation frequency per electrode in real time. Arm movement is activated by the recorded neural activity, using averaged firing rates.

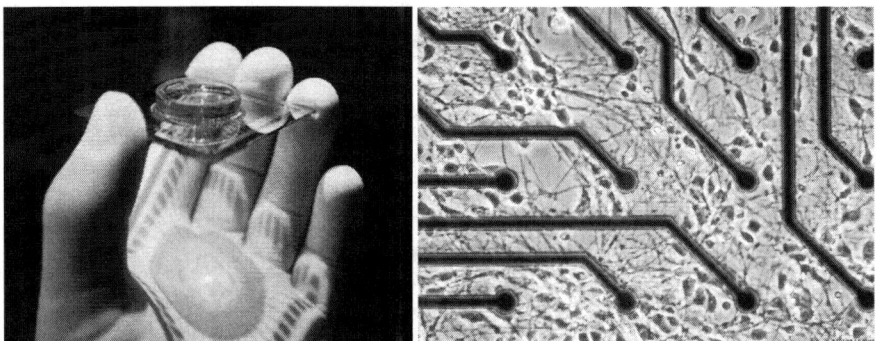

Figure 3: MEART, the Multi Electrode Array and the neuron culture.

However, the drawings resulting from this activity do not seem to follow a representational programme: they do not aim at producing a likeness but rather visual traces of the neural activity. Their transcription of the video portraits being sent to the brain leaves what look like unpredictable traces. The installation at the exhibition venue shows the drawing robot in action, with videos of the neural culture in real time, along with mappings of its activity and diagrams of the image evaluation process. The public can experience the drawing activity and at the same time sense the presence of its remote initiator, the brain. This phenomenological experience gives the sensation of a fragmented network body. Sometimes, accidents occur: the upper limit of marks on the paper is reached and the sheet is torn. At other times, the drawing activity may stop. The lab sends a message saying that the neural culture has just died.

Another version of this work, called Silent Barrage, was first shown in New York in 2009. The overall process is the same but the robotic body becomes an immersive space the public can walk around. The gallery space is mapped by several cameras mounted on the ceiling and directed at the floor. It is filled with a number of vertical poles fitted with a drawing device that leaves marks on their cylindrical bodies. Each pole is connected to a specific region of the remote brain. The visitors' movements are recorded by the cameras and analysed using a position-mapping program. This information is transmitted to the Multi Electrode Array where the neuron culture responds to it, stimulating the drawing devices on the poles. A loop is thus established between the gallery space, the spectators, the brain and the drawings. Walking into the installation is a way of initiating an interaction with the remote brain that becomes noticeable through the accumulation of marks being inscribed on the poles in real time.

Figure 4: Silent Barrage, 2009, view of the installation at Exit Art, New York.

Compared with the two-dimensional drawing process of MEART, this immersive drawing machine marks a spatial change. Until now, the Potter's Lab scientists have only obtained two-dimensional monolayer cultured neurons. However, functional importance lies in the possibilities given by the three-dimensional layered nature of the cortex and scientists are now pursuing the construction of 3D Multi Electrode Arrays to support three-dimensional neuron cultures. When designing the Silent Barrage installation, the artists somehow metaphorically materialized what the scientists are seeking to achieve. My own interpretation is that these drawing columns may echo the 'cortical columns' that are so important in the development of neuronal systems. Their extra long axons allow neurons from different cortical layers to communicate.

Where does Silent Barrage's poetical title come from? In a paper entitled 'Removing some "A" from AI: Embodied cultured Networks', members of the interdisciplinary team report what they call 'barrages' of neural activity spreading at a high firing rate throughout the entire neural system, when the in-vitro culture is not connected to afferent inputs:

> Barrages… are continuous over the life of the [in-vitro] culture. We consider the possibility that [they are] a consequence of sensory deprivation… If more than 30% of the neurons are endogenously active, the neurons fire at a low steady rate… We are developing… mappings in which continuous sensory input quiet barrages, bringing the networks to a less 'sensory-deprived' state that allows more complex, localized activity patterns. (Bakkum *et al.* 2004)

'Silent barrage' is a poetical formulation of the fact that the neural system's spontaneous activity, involving permanent firing, may be calmed down and organized by sensorimotor interaction.

Dr. Steve Potter's Lab, where the neurons are cultured, is part of a neuroengineering laboratory that creates neurally controlled biobots in order to investigate ways of reducing the distance between artificial and biological intelligence. In these entities, biological in-vitro neurons are connected to robots through Multi Electrode Arrays. The in-vitro culture enables careful observation of neural activity, while the interaction with robots leads to specific developments of this activity in terms of memory, learning, agency and intelligence. These experiments are designed to observe how the biobot behaves while handling specific tasks. Complex animal behaviours emerge and develop from the interaction between an organism and an environment. The Potter's Lab scientists analyse the firing patterns of neurons confronted with inter-stimulus intervals that vary as a situation develops. For instance, a robot placed under neural control might successfully approach a moving target and maintain a constant distance from it. Learning processes are initiated, using mechanisms of neural plasticity.

However, with MEART or Silent Barrage, the sensorimotor mappings are less well defined. The drawings change throughout the life of the neural cultures, and vary according to different cultures and to the variations in the audience. The robot's behavioural response sheds light on the properties of the neural network and directs further encoding refinements. Dynamical systems theory may help build models of embodied cognition, according to the authors of the above-quoted paper:

> These processes can be expressed using dynamical systems theory (DST), a mathematical framework to describe systems that change in time... DST contends that multiple feedback loops and transmission delays, both of which are widespread in the brain, provide a time dimension to allow higher-level cognition to emerge without the need for symbolic processing. DST is a framework compatible with embodied perspectives. (Bakkum *et al.* 2004)

But these experiments are as yet somewhat rudimentary, and Silent Barrage is a work in progress that raises more questions about the relationship between neural mechanisms and creative consciousness than it answers. Can this artwork be linked to the well-known philosophical 'brain in the vat' thought experiment based on the hypothesis that we are all disembodied brains floating in a vat of nutrient fluids, as in the movie Matrix? Here the brain is connected to a computer running complex programs that send electrical stimuli through sophisticated networks of electrodes, so as to mimic how normal brains are stimulated as a result of perceiving external objects and interacting with them. The philosophical question raised by this thought experiment is: granted — following the Cartesian *cogito* — that we can be certain of our own existence from the moment we are conscious of our own consciousness, how can we assert that the real world also exists outside of our consciousness and mental representations? By communicating through representations and language, the brain in the vat does sense the real world and share what it takes to be its own experience of it, just as a normal embodied brain does. This supposition is close to Descartes' 'Evil Genius' hypothesis where the Genius takes on the role of the computer and deceives us by inducing false beliefs about the outside world (Descartes 2008, First Meditation). But how can we prove that the real world does exist and is not just a neuronal movie taking place in our brain?

In a well-known demonstration, Hilary Putnam attempts to refute the sceptical claim that this is impossible to prove. The philosopher bases his argument on semantic externalism, which can be summed up by his famous 'meanings just aren't in the head' (Putnam 1982). To cut a long story short, the difference between the representations formed by an envatted brain and those produced by a normal embodied brain—despite their being qualitatively identical as a perceived phenomenon—lies in the fact that the former refers not to real things but merely to images of real things, whereas the latter refers to a direct personal experience of the real world. The representations may be similar on a phenomenological level, but on a conceptual level the references are radically different. The occurrence of a mental representation, relying on words or images, cannot be equated with the ability to refer to the real world by conceptual means. All mental representations do not automatically refer to a referent.

MEART's creators have in a way turned the 'brain in the vat' problem upside down. They do not show us the behaviour of a fully-grown disembodied brain reacting to stimuli that exactly reproduce perceptions of the real world. They have designed their experiment to observe how, starting from a few cells, a neuron culture develops an agency when confronted with a sensory-motor situation through an electronic interface, which is also connected with devices allowing real physical action and perception. Instead of seeking to prove, as the 'brain in the vat' thought experiment does, that the mental representations produced by stimuli originated in a mature brain by a digital program are equivalent to representations referring to the real thing, they tend to show how a developing intelligent system constructs its representation and mapping capabilities by interacting with external elements. Without this interaction, the neurons would not develop localized patterns of activity but would instead produce overall noise by firing indiscriminately ('barrage').

Of course, there can be a problem with the kind of programs used in the digital interface to translate the visual images coming from the exhibition space into electrical stimuli entering the brain, and to transpose the reactions of firing neurons into mechanical movements to produce drawings. Our sensory-motor system is analog, not digital, whereas MEART is a combination of both modes, which in a way makes it an envatted brain. It is not a matter of ascertaining whether MEART constructs its capacity of agency by interacting with its environment; this has already been proven by the elaboration of evolving patterns that are visible in the electronic mapping of the Multi Electrode Array (electric patterns) and in the drawings (graphic patterns). It is a matter of defining the kind of representations that help MEART build these patterns. Is what these emerging and ephemeral representations refer to a human portrait, the shape of an animated being, or is it not rather sixty areas of grayscale values in a given format? Does MEART's agency involve intentions and even some kind of consciousness? Can we describe this semi-living artist as being an emerging subject? Last but not least, do these questions make any sense in this context?

MEART's creators have produced a cyborg by mixing biological and mechanical elements and by giving them agency through analogical and digital translation modes. As with Alexitimia, which involved fewer elements, MEART is a nexus connecting the brain, the computer program, the Internet, the gallery space, the drawing arms, the sheet of paper, the camera and the viewers. It is a situated process, not geographically situated, since the organs of its body can be spread wide apart, but where each element plays a role within a loop formed by the interaction between the brain and the audience. The visitor is conscious of the presence of this other, while the brain observes the public and reacts to their movements (more specifically in Silent Barrage). Continuity is established between externality and interiority.

What is the 'subject' of this experiment? Is it the audience or the brain? The work of art, here, is not an object but a situation involving at least two subjects looking at each other (the brain and the public). Here, in a very clear manner, the subject of the experiment is also part of its content. MEART is an interplay of actions, perceptions, reactions and corrections producing traces. It is a semi-living system of distributed agency, a 'smeared out biotechnological matrix' (Clark 2008, p. 80) that goes to make up an entity comprising several subjects. This semi-living artist's creative mind is a distributed entity, and so is the creative mind of the visitor participating in the piece.

Both artworks I have presented possess an aesthetic effectiveness lying in the phenomenological presence of a cyborgian 'other' rather than in any symbolic demonstration. As Paul Vanouse writes in his essay, 'Contemplating MEART — the semi-living artist': 'MEART's creators have cleverly designed their thought-provoking apparatus to maximize cognitive dissonance' (Vanouse 2006). This 'cognitive dissonance' may reflect the gap existing between the neurons' mechanical activity studied by neuroscientists and consciousness as we can experience it in an aesthetic experience for instance. I think that the 'cognitive dissonance', which may be experienced in a work of art, is a prerequisite for an aesthetic experience. Like empathy, this kind of experience is linked to neurological mechanisms, but its 'dissonance' shows that it does not just follow *a priori* whether neurobiological or aesthetic rules.

We would probably all agree that an artwork has to do something more than just illustrate a scientific demonstration or apply a scientific law. Furthermore, whatever the media it is embodied in, an artwork is not reducible to its most basic components. This reduction certainly helps us to analyse the way we perceive things in general, and artworks in particular, and also to understand our sensorial mechanisms. But a painting, for instance, is not just so many basic visual elements successfully combined on a canvas so as to produce a complex, attractive and pleasurable ensemble. This definition was heavily criticized by Marcel Duchamp more than half a century ago as being what he termed 'retinal painting'. He did not mean that art should not be experienced through sensory perception, or even that it should not give sensual pleasure, but that there is another dimension to it that makes it art. This dimension is connected with the artist's intention and not only with its ability to materialize visual relations. It places its viewers in a specific position, simultaneously driven by their aesthetic feeling and actively completing the art process. A substantial part of aesthetics has developed around this idea.

MEART-Silent Barrage may be seen as a sophisticated toy invented by artists and scientists to play with together. The public may also experience it as a lively environment that fosters communicating and addressing neuroscientific research issues. But does that make it an artwork? When the artist and writer Paul Vanouse entitles his essay 'Contemplating MEART', he is referring to something more than playful communication. Contemplation is an active immersion, which is somewhat paradoxical since it requires letting go of your consciousness in such a way that you feel you are entering whatever you are contemplating, or it is entering you, and the distinction no longer exists between you as a subject and the object of your contemplation. As a state of consciousness, active contemplation is probably related to empathy. Its neurobiological basis is to be found in the architecture of the neural system, as a kind of internal form echoed in perceived reality. However, I would like to take issue with this viewpoint since it may lead to the statement that art is a representation of the essence of reality.

4 Secretion and creative flux

This pre-existing 'essence' is to be found in the correspondence between the outside world and our receptive apparatus, including our neural architectures. But this conception may turn the cognitive and even the aesthetic experience into the processing of information according to programs following set *a priori* cognitive rules, and this is not the path I am following in seeking to reconcile brain, body and world without including them in any pre-existing scheme. It is important here to reject the premise of an *a priori* essence of reality that the subject might perceive and understand through a mostly hermeneutic relation.

On the contrary, I support the idea that the relationship between an organism and its environment is more like a field or flux of common properties blurring the boundary between the subject and the object. This view is connected to J.J. Gibson's notion of 'affordances'. For the psychologist, affordances of a given entity form 'a specific combination of the properties of its substance and its surface taken with reference to an animal' (Gibson 1977, p. 77). Affordances distributed in an environment offer the tools for an animal organism to effectively incorporate its environment, build its receptive fields and develop its ability for agency. Affordances may combine with symbolic thinking but can be perceived without it. For instance, the affordances of a doorway to move to the other side of a partition are first perceived through visual or tactile stimuli with information about its size in relation to the size of the user's body (Warren and Whang 1987). This is not to minimize the symbolic dimension of the doorway, on the contrary. Affordances are neither totally in the object nor in the subject, they are an in-between, stimulating subjective behaviour while contributing to an objective function. They are more of a flux between two poles — the subject and the object — which do not really exist *per se*, in their extreme positions, but which, among other properties, really come alive through the mediation offered by these affordances.

We will now approach the theory of enaction, which makes even more obsolete the border between a subject founded on its own interiority and the world of objects. With its constant and multiple loops between perception, stimuli and reactions, and with its somewhat ambiguous mapping program instead of the more task-oriented experiments usually carried out at Potter's Lab, MEART shows us an 'enaction' process, as Francisco Varela has described it: the spontaneous growth and organization of a complex system interacting with its environment. 'Autopoiesis', or spontaneous growth, to quote the term coined by Humberto Maturana, is achieved through an endogenous activity of the system coupled with its exogenous activity, and this coupling is used by the cognitive process to bring about a coherent world (Maturana and Varela 1980; 1987). In their book *The Embodied Mind*, Francisco Varela and co-authors question neuroscientific protocols:

> In general, within enactive cognitive science, a process akin to evolution as a natural drift takes the place of task-oriented design. For example, simulations of prolonged histories of coupling with various evolutionary strategies enable us to discover trends wherein cognitive performances arise. (Varela *et al.* 1991, p. 207)

This vision has been passed on by more recent philosophers of consciousness such as Alva Noë (Noë 2004), or by the neuroconstructivist theory related to developmental and evolutionary robotics (Mareschal *et al.* 2007). An enactive approach criticizes the notion of 'representation' as it is commonly used:

> The key point is that such systems do not operate by representation. Instead of representing an independent world, they enact a world as a domain of distinctions that is inseparable from the structure embodied by the cognitive system. (Varela *et al.* 1991, p. 140)

From an enactive perspective, a cognitive performance is not an efficient representation of a pre-existing real world. On the contrary, it is the creative emergence of a world through the coupling of the endogenous and exogenous activities of a complex system. Of course, a cognitive process is based on various sorts of representations, but these representations are emerging and evolving images rather than fixed architectural frames. Neuroscientists are focusing more and more on distributed fragmented representations. The story that MEART is telling the public—the resultant drawings are only a part of it—is in the process of being written as it is being enacted. Rather than perfecting a specific task, it is undergoing a 'natural drift' as Varela puts it. Instead of an intelligent process responding to pre-established criteria, what the public sees in the drawings is the noise produced by the working neurons and the entire nervous system, which is shaped by the brain, the electronic interfaces and the Internet. This noise may be a manifestation of the construction of intelligence—and the drawings produced by the robot indeed show common patterns, with variations from one piece to another, indicating a form of cognitive elaboration. It is even possible to place an aesthetic value on these graphic works (they are exhibited and sold simultaneously as traces produced by an artwork—the semi-living artist—and as works of art *per se*), but it is difficult to situate them in the field of visual productions somewhere between, for instance, the rhythm of an electrocardiogram and Cy Twombly's magic scribbling. In Shannon's theory, information communication systems produce symbolic meaning against a background of 'noise'. Meaning and noise are not antagonistic. Meaning emerges from noise. A secretion of noise can become a projection of meaning and *vice versa*.

According to Francisco Varela, autonomous biological systems, like neural networks, can spontaneously organize themselves, depending on a condition of 'operational enclosure'. This enclosure is the delineation of a system becoming distinct from its environment and it could become a problem in a theory of fluxes and continuity between interiority and consciousness on the one hand and external reality on the other. However, as we have already seen, the principle of delineation does not imply radical enclosure, just as skin, which is the limit of the body, is also a porous interface. The term 'enclosure' suggests that there is a gap between a consciousness and a world; however, this gap mostly shows that the relationship between such a system and the world is not merely a question of causality. Consciousness is not the mechanical result produced by a set of stimuli initiated by a ready-made reality and processed by a cognitive system, even in a fragmented way, as if through a broken mirror. It is rather a way of enacting the world within a large baroque theatre. This is not to say that the world is a mere solipsism, nor does it mean that consciousness is alone in its enclosure. The anxiety of being alone comes from a certain conception of the duality between the subject and the object. It appears with the need for an objective evaluation of the world from a remote viewpoint, which should be as immutable and universal as possible — the viewpoint of a *cogito*, of a subject grounded in its quest for objectivity. However, consciousness is not alone and the world is not an illusionary projection, since consciousness is the world. According to Varela, 'organism and environment enfold into each other and unfold from one another...' (Varela *et al.* 1991, p. 217). Consciousness is surrounded by other beings, like a fold within larger folds. It is like the world, on another scale, but within a fractal system rather than as a microcosm reflecting the macrocosm. Membranes, which are among the main factors of enclosure, become interfaces, surfaces of contact and exchange between the brain and different worlds. By synthesizing this relationship, membranes become metonymies for the world as for the embodied consciousness, echoing Antonin Artaud's vision, which I have put at the beginning to this paper: 'Man... like a skin that walks... buckling on one's own humanity'.

We are now moving away from a model based on representation — the world as the projection and deciphering of a given reality in a neuronal movie theatre — towards an enaction model, a perceiving and acting subject embodied in a world, at once shaping it and being shaped by it. Perception is not only a stimulus to enter but also the act of building a world of phenomena. Maurice Merleau-Ponty puts it this way:

> Thus, the form of the excitant is created by the organism itself, by its proper manner of offering itself to actions from the outside... 'The environment (Umwelt) emerges from the world through the actualization or the being of the organism — [granted that] an organism can exist only if it succeeds in finding in the world an adequate environment.'

(Merleau-Ponty 1963, p. 13; the quotation is from Kurt Goldstein 1963, p. 88)

A few pages later, the philosopher adds: 'The relations between the organism and its milieu are not relations of linear causality but of circular causality' (Merleau-Ponty 1963, p. 15). Therefore, biological evolution cannot be reduced to maximum efficiency in the relationship between an organism and its environment. Adaptation and natural selection tend to diversify the plasticity of the living through interferences, rather than maximize a system and upgrade it according to pre-established schemes. This plasticity of biological systems blends with de-territorialized cognitive models such as networks that would weave a sort of peripheral intelligence. Varela coined the expression 'natural drift' for this conception of the evolution of the living: 'Evolution as a natural drift is the biological counterpart of cognition as embodied action...' (Varela et al. 1991, p. 188).

A 'natural drift' involves accidents, abnormality and dissonance in its creative movement. I support the view that art is a way to perceive cognitive dissonance as a creative process. It makes us witness a cognitive drift rather than a model-based development. How does that cognitive drift become a shared sensibility, a sensibility embodied in artworks that may address not only their contemporaries but other times and cultures as well? Does the existence of neural systems that are specialized in rewarding and active in empathy explain all the complex process of pleasure that we experience with art? I think that, due to their psychological and socio-cultural dimensions, aesthetic experiences are more complex, especially since they are a way of feeling the consciousness we have of our own consciousness.

But what is consciousness? Referring to a theory of distributed fragmented consciousness developed by Semir Zeki and his colleagues (2003), Riccardo Manzotti makes this statement:

> There are many microconsciousnesses distributed in space and in time each corresponding to the occurrence of a specific phenomenal content... Each microconsciousness can be seen as the neural part of a larger process — beginning in the external world — whose neural counterpart is the microconsciousness. (Manzotti 2006)

This continuous flux between the external world and the internal psychic activity has been coined by W. Teed Rockwell as a 'brain-body-world nexus' (Rockwell 2005, p. 71). In my opinion, this theory ties in rather neatly with Gilles Deleuze's theory of 'the fold', where microperceptions form a folded continuous fabric that binds together perception (matter) and consciousness (mind), which are the two levels of Leibniz's monad, according to Deleuze (1993). It forms a flux in which microperceptions construct larger patterns, as smaller inflexions are enfolded in larger ones. In that system, the mind is not merely perceiving the world at a distance but actualizing it in a simultaneous process. At the end of the book,

Deleuze adds that the enclosed monad conceived during the Baroque period and whose outside referent was God is now converted into an opened monad directly connected with the outside world.

5 A secretory perspective

These theories tend to blur the distinction between the subject and the object, and especially between the viewer and what is perceived. A projective model relies on the kind of gaze which is able to focus on and aim at things: it would not be feasible without a precise viewpoint from which the action of pinpointing becomes possible. The viewpoint is the place of the subject, while what is pinpointed is the object. This kind of predator's gaze tends to create a distant and objective confrontation, described by Rosalind Krauss as

> ...the operation of the model by which subject and object are put into reciprocity as two poles of unification: the unified ego at one end and its object at the other. Lacan had called this model 'geometral,' and had identified its rules of perspective with the assumptions grounding the Cartesian subject. (Krauss 1996, p. 96)

This model is based on a central viewpoint that is both immersed in the observed world, and distant from it, so as to look at it as through a window. As a method of representation, it has produced the conical projection of perspective, for instance, which is linked to the advent of the modern subject. But there are many other ways to look at worlds, and many other artistic strategies and aesthetic regimes to express them. A nomadic gaze is more diffuse or absorbent, functioning as a secretory mode of interaction, drifting between things rather than aiming at them. According to Rosalind Krauss, instead of a fixed geometrical point, the subject becomes a 'stain' whose delineation in its environment fluctuates:

> But the Gaze, as an irradiant surround, comes at the subject from all sides, producing the subject now as a stain rather than a cogito, a stain that maps itself... onto the world's 'picture,' spreading into it, getting lost in it, becoming a function of it, like so much camouflage. (*ibid.*, p. 96)

Cognition is no longer based on the representation of a world by a subject processing information in order to build a map of that world. As a stain spreading over a background, the subject is mapped onto the world's surface. There are multiple viewpoints coming 'from all sides' and their cohesion is diluted in the 'irradiant surround'. Their common delineation is porous.

The secretory model presented in this paper functions more like a contamination, a dissemination and an interference than a projective representation, like a drift rather than a programmed evolution. It works by hybridizing rather than by purifying. It is quite difficult to devise scientific experiments to test the hypotheses this model is based on and even more so to build a cognitive model of 'cognitive dissonance'. From this standpoint, the dialogue between art and aesthetics on the one hand, and biology on the other, is quite promising, provided it is a genuine two-way process. The hard part is upholding the purity or distinctiveness of each discipline, while effecting their mutual contamination!

Thanks to the cutaneous interface, we have put the brain in the body, and the body in the world, constructing the notion of brain-world. In art history, the rare presence of the motif of the flayed skin, which has become autonomous from the rest of its body and takes on a life of its own, is a hint in the direction of the notion of the brain-world. We can now observe numerous symptoms of its presence in the technological interfaces that surround our everyday life, such as computer screens, which are built on a projective model. Let us be aware of the danger that these hints of brain-world may easily drift towards a more and more dematerialized, abstract dimension (a reaction to this is to be found in the development of tactile screens, for instance). Our secretory operative modes, relying on embodiment, are vital to our creative capacity.

Acknowledgments

Many thanks to Fiona Reverdy and John Lee for their corrections of the English text.

All illustrations courtesy of the artists.

References

Aelian (1997), *Various History*, 13, 21, D. Ostrom Johnson, Trans., London and New York, Edwin Mellen Press.
Artaud, A. (1976a), 'The Peyote Rite in Tarahumara Country' in *The Peyote Dance*, H. Weaver, Trans., New York, Farrar, Straus & Giroux.
Artaud, A. (1976b), 'To Have Done with the Judgement of God', a radio play, 1947, in S. Sontag, Ed., *Selected Writings*, H. Weaver, Trans., Berkeley, University of California Press.
Bakkum, D.J., A.C. Shkolnik, G. Ben-Ary, P. Gamblen, T.D. DeMarse and S.M. Potter (2004), 'Removing the "A" from AI: Embodied Cultured Networks' in L. Steels and R Pfeiffer, Eds., *Proceedings of the Dagstuhl Conference on Embodied Artificial Intelligence*, Heidelberg and New York, Springer. Also [Online], http://www.fishandchips.uwa.edu.au
Changeux, J.P. (1997), *Neuronal Man: The Biology of Mind*, L. Garey, Trans., Princeton, Princeton University Press.
Changeux, J.P. (2002), *Raison et Plaisir*, Paris, Odile Jacob.
Changeux, J.P. (2008), *Du Vrai, du Beau, du Bien: Une Nouvelle Approche Neuronale*, Paris, Odile Jacob.
Clark, A. (2008), *Supersizing the Mind: Embodiment, Action, and Cognitive Extension*, Oxford, Oxford University Press.
Connor, S. (2004), *The Book of Skin*, London, Reaktion Books.
Deleuze, G. (1993), *The Fold: Leibniz and the Baroque*, T. Conley, Trans., Minneapolis, University of Minnesota Press.

Deleuze, G. and F. Guattari (1987/2004), 'November 28, 1947: How Do You Make Yourself a Body without Organs?' in *A Thousand Plateaus, Capitalism and Schizophrenia*, B. Massumi, Trans., London and New York, Continuum.

Descartes, R. (2008), *Meditations on First Philosophy*, M. Moriarty, Trans., Oxford, Oxford University Press.

Dumas, S. (2008), 'The Return of Marsyas: Creative Skin' in J. Hauser, Ed., *SK-Interfaces*, J. Lee, Trans., Liverpool, Liverpool University Press. A previous version of this paper was published in German: (2006), 'Der Mythos des Marsyas, ein Bild Paradigma' in U. Renner and M. Schneider, Eds., *Häutung. Lesarten des Marsyas*, W. Kukulies, Trans., Munich, Wilhelm Fink. Also [Online], http://www.stephanedumas.net

Gibson, J.J. (1977), 'The Theory of Affordances' in R.E. Shaw and J. Bransford, Eds., *Perceiving, Acting, and Knowing: Toward an Ecological Psychology*, Hillsdale (NJ), Lawrence Erlbaum Associates.

Goldstein, K. (1963), *The Organism: A Holistic Approach to Biology Derived from Pathological Data in Man*, Boston, Beacon Press.

Homer (1914), 'To Apollo', *Homeric Hymns*, 131, H.G. Evelyn-White, Trans., Cambridge (MA), Harvard University Press; London, William Heinemann Ltd.

Homer (1924), *The Iliad*, I, 14, A.T. Murray, Trans., Cambridge (MA), Harvard University Press; London, William Heinemann Ltd.

Krauss, R. (1996), 'The Destiny of the Informe', *OCTOBER*, **78** (Fall). Also in Y.A. Bois and R. Krauss (1997), *Formless: A User's Guide*, New York, Zone Books.

Manzotti, R. (2006), 'A Process Oriented View of Conscious Perception', *Journal of Consciousness Studies*, **13** (6): 7–41.

Mareschal, D., M.H. Johnson, S. Sirois, M.W. Spratling, M.S.C. Thomas and G. Westermann (2007), *Neuroconstructivism: How the Brain Constructs Cognition*, Oxford, Oxford University Press.

Maturana, H. and F. Varela (1980), 'Autopoiesis and Cognition: The Realization of the Living' in R. Cohen and M. Wartofsky, Eds., *Boston Studies in the Philosophy of Science*, vol. 42, Dordecht, D. Reidel Publishing Co.

Maturana, H. and F. Varela (1987), *The Tree of Knowledge: The Biological Roots of Human Understanding*, Boston, Shambhala.

Merleau-Ponty, M. (1963), *The Structure of Behavior*, A.M. Fisher, Trans., Boston, Beacon Press.

Noë, A. (2004), *Action in Perception*, Cambridge (MA), MIT Press.

Putnam, H. (1982), 'Brains in a Vat' in *Reason, Truth, and History*, Cambridge, Cambridge University Press.

Ramachandran, V.S. (2005), 'The Neurological Basis of Artistic Universals', *Art and Cognition Workshops*, Paris, Institut Jean Nicod, EHESS/Ecole Normale Supérieure, [Online], http://www.interdiscipines.org/artcog/papers/9

Ramachandran, V.S. (2007), *The Artful Brain*, New York, PI Press.

Rizzolatti, G. and C. Sinigaglia (2008), *Mirrors in the Brain: How We Share our Actions and Emotions*, Oxford, Oxford University Press.

Rockwell, W.T. (2005), *Neither Ghost nor Brain*, Cambridge (MA), MIT Press.

Vanouse, P. (2006), 'Contemplating "Meart—The Semi Living Artist"', [Online], http://www.fishandchips.uwa.edu.au

Varela, F., E. Thompson, and E. Rosch (1991), *The Embodied Mind, Cognitive Science and Human Experience*, Boston, MIT Press.

Warren, W.H., Jr. and S. Whang (1987), 'Visual Guidance of Walking Through Apertures: Body-Scaled Information for Affordances', *Journal of Experimental Psychology: Human Perception and Performance*, **13**: 371–3.

Zeki, S. (2000), *Inner Vision: An Exploration of Art and the Brain*, Oxford, Oxford University Press.

Zeki, S. (2003), 'The Disunity of Consciousness', *Trends in Cognitive Sciences*, **7** (5): 214–8.

Figure 1: Galetta's Exhibition — 'Il Museo del Caos'. Museo di Arte Contemporanea di Villa Croce, Genoa, 2010

Giuliano Galletta

The Self is Around Me

Giuliano Galletta is a renowned Italian artist whose work is exceptionally close to the gist of the present collection of essays. Since this is a volume on the relation between a model of the mind and art, it seems fair to host a lively and personal account of how much art, mind, personal life, and world become seamlessly intermingled in the current intellectual and artistic European milieu. Galletta is well known for his many artistic and public alter egos (Leo Sarastro is one of them) that he has been painstakingly building out of scattered and publicly available events, snapshots, drawings, and artistic events. Galletta is the human being whilst his alteregos are public and distributed entities insatiably swallowing their author's private and public life. The following is not to be read as an academic essay but as the first-person testimony of an artist devoted to unfolding the structure of the subject according to an externalized perspective. Rummaging and rumbling at many epistemic and aesthetic levels, Galletta gets to one of his favourite epistemic and aesthetic devices: the almanac. For him the almanac is a way to externalize a tentative self. If a self is made by its content, any scattered collection of materials might define a self. By means of comparing the fictitious collection of memory traces (tickets, shots, cards, newspaper cuts, and the like) to a real memory, he challenges the traditional model of the self. He sets aside worn-out boundaries. Like other authors (he quotes unabashedly Wallace Stevens), he targets the received view of the self. His goal is to show the emptiness of the myth of an internal self as it is usually concocted by common sense and by the society. [Editor's note]

> *Say what you will, twelve thoughts, each of a single word, are not the self-same mental thing as one thought of the whole sentence.*
> William James (1909)

Everyone's life is built day by day. Of course, some people strive to plan and control every detail. Some parents plan their children's life since birth, and some children would like to control their parents' final days. Most of these attempts are doomed to fail, especially those that seem to have the greatest chance of success. These are examples of a hubris for the control of the life of subjects that is also a desire for its destruction. Yet, notwithstanding such efforts, no individual succeeds completely in achieving both self-destruction and destruction of other selves because any person is akin to a collage whose pieces are singled out by greater forces such as world and destiny.

The *self* I am referring to—shaped by our western and capitalistic civilization—is the unavoidable and undesirable result of an historical and anthropological evolution. In different cultures, there might have been totally different kinds of *selves*; yet, I believe the selves we face now have been more or less the same for the last few centuries.

The *self*, which I believe to be the most cumbersome of all the nouns, is unstable and weak. The self can almost be seen as a serious clinical condition because it hampers subjects from integrating with a rule-based society. To cope with society, individuals try to overcome their weakness by using a large amount of narcissism. This social 'therapy' is systematically spread through the media, mostly by television and advertisements, but also via more sophisticated cultural instruments such as cinema, literature, psychology, arts, religion and politics. Narcissism helps the self to survive in society.

In my work, I challenge the substantiality of the self both as it is conceived by the layman and as it has been concocted by the prevailing western cultural *milieu*. On the basis of their status and background, everyone is encouraged and helped to accomplish the emphatic imperative 'be yourself'—an imperative which is trivial and impossible at the same time. The tools provided or sold to achieve this goal are the same for everyone, albeit disguised differently for everyone. The psyche (or personal identity) is a tool created purposefully to prevent the achievement of the goal of being oneself. They prevent everyone from being himself/herself. The same happens by asking to be spontaneous—another paradoxical injunction. The self is one's identity hallmark and key. In other words, the belief in the self prevents one from being what one really is. It is a self defeating hypothesis.

Such a paradoxical inconsistency is the cause of neurosis. For society treats us as if we were 'non-entities' whilst, at the same time, we are expected to be 'somebody'. I do not know if it is possible to avoid such a mechanism by simply rejecting the paradoxical injunction and then refusing to be 'ourselves', 'spontaneous' and ignoring the brief moment of glory Andy Warhol entitled us to have.

Metaphorically speaking, to steer away from the 'Cyclops', epitomizing the controlled world concocted by an alleged community of selves, we ought to give Ulysses' answer: 'I am nobody'. But who dares to be brave enough? Would Ulysses' crew—mentioned by Adorno and Horkheimer—

have been able to show the same prowess as their master? Getting rid of the false promises of the notion of the self is not an easy feat.

In literature, the deconstruction of the self is an easy and often-played game. Consider Luigi Pirandello's 'One, No one, and One Hundred Thousand'. The author's feel of estrangement from reality helps a great deal. In everyday life, playing with the self is much more difficult and dangerous. Luckily, not as dangerous as a hypothetical reinforcement of rules based on identity and its isolation. The everyday self is allegedly floating over an abysmal isolation. How to conceal such a precept with the fact that our existence is a do-it-yourself project whereas the *self* becomes the *other* or even the *others* as in Pirandello's works?

A viable dictum to get rid of the received view of the self is the apparently paradoxical 'I don't want to be me'. Lovers sometimes perilously venture to say: 'I would like to be you', without knowing the risks.

The *other* is not just another individual: it could be a whole community or an abstract social layer. The *other* might be a person but also a part of nature, an animal, or a thing. It might become 'a thing among things' (Rilke 1974). It might be an object stolen from today's maniacal world of consumption—a transfigured phantasmagorical good.

Yet the self does not exist in isolation. It does entail the other. Thus it entails being with the other—a condition that doesn't necessarily imply having a story to tell; nor does it prevent anyone from telling a story, though. Consider a form of self creation: writing a diary. It is an impossible task for a lover, and not only for him, as he cannot explain what he's feeling. Roland Barthes (2003) writes:

> We must, no doubt, conclude that I can rescue the Journal on the one condition that I labour it to death, to the end of an extreme exhaustion, like a virtually impossible Text: a labour at whose end it is indeed possible that the Journal thus kept no longer resembles a Journal at all.

Paradoxically, the ideal case for writing an autobiography is not having a biography at all—obviously a condition impossible to obtain. Whoever has a biography is the main character of facts and events which are memories gathered into a personal mythology. This is a narration allegedly trying to coalesce facts and events into a self—a 'novel' that unfortunately does not always find an author like Proust eager to frame it. The self self-defeats itself, as to prove its self-inconsistency. What is the hallmark of a 'distinguished', 'interesting' or 'adventurous' life worth of being told? Any sub-culture, any form of art either professional or even infantile, any collection of personal items, any family album, teaches us that there are several steps involved, from the willingness to tell a story to sharpening the skills necessary to craft a narrative world. The author of a biographical narrative chooses to follow a perilously intimate road—a faintly marked path containing all of these concepts in an extremely epigonic way.

My goal is to write a 'lifeless' autobiography and not simply a literary work where a character says 'I'. I aim at producing a true 'life' through a 'false' tale, possibly with the undeclared purpose of showing that life is never totally 'true'. In order to do this, I followed the path of the great avant-garde novels of the nineteenth century, which have systematically influenced other genres and have created confusion between art and life. Art is based on the explosion of the substantial and essentialist self of the modern world. The language and style used do not adhere to those utopias that current society has already managed to achieve.

According to Edoardo Sanguineti, every act of creativity springs from a difficult situation, either from an absence or from an obstacle. Telling a story without anything to say is a macroscopic mistake. Creating a *self by means of* putting together other scattered *selves* is akin to stitching a Harlequin dress.

The avant-garde novels championed the power of Chance and Chaos as means to rebel against Tradition. In turn, the postmodern Society of the Spectacle encourages consumption and turns Chance and Chaos into the paradigm of everyday life. In 1947, Adorno could still say that 'the task of art is to introduce chaos into order' (Adorno 1954). Karl Kraus (1987) echoed that 'chaos should replace order as the latter has not worked'. Currently, the problem is just the opposite: is it possible to rebuild a narrative order that contains Chaos and takes advantage of it? My hunch is that the answer lies in the relationship between word and image, between object and behaviour.

In 2005, trying to define a method, Galletta drew an open circle on an A4 size paper with an indelible felt-tip pen. Then, starting from the left side and in a precise order, Galletta wrote these words: document, action, image, object, situation, body, environment, world, text, story, and scenery. The drawing was entitled 'Inspiration Plan'. Its description was:

> The anti-biography is one of the numerous techniques of auto-anthropology in which reality turns into dream. Memory becomes amnesia, truth becomes an error. In order to achieve these results, it is necessary to resort to Chance (and to Chaos). However, using Chance (as well as Chaos) for our own aims it is not an everyday-accomplishment and it takes an intense activity that some individuals, since past centuries, have decided to call with an obsolete term: inspiration. It is a series of simple procedures that can be summarized in this scheme.

Earlier in a previous text ('Auto-presentation') in 1981, Galletta explained his methods in unambiguous terms:

> I have restrained myself from going through all the impossible circumstances of my existence. So I have written another biography. The experiences I have gathered concern a brief period, the right amount of time needed to 'objectify' them. On the eleventh day, I found myself working on these small, private aesthetic events, a sort of organized disinformation on a set space. This work is confined within the idea of

page: it doesn't escape from it and it couldn't anyway. It is just another small, irrelevant crumbling idea no more important than the roughness of the material used, with scarce psychological meaning. An expert may easily recognize the psychopathological symptoms of the obsessive condition, by means of symptoms that are normally ignored insofar as they do not concern us. Systematically, we forget, by means of appropriate cultural updates, what we have never learnt. We treasure the memories of our school education and witness the tension between what we ought not to have said and what we ought to have done. Sexual perversions are part of these memories. I will skip their analysis although their cultural value cannot be underestimated.

It might be useful to point out that I wrote the above text — with the apparent aim of 'presenting' a certain 'aesthetic' artefact — as a virtual introduction to something that was yet there and that might have never existed. Another related text, conceived as a self-introduction, was written in 2003 on the basis of the same logic by an author under the pseudonym of Leo Sarastro:

> Even 'non-facts' leave traces. This almanac is built on such clues. If we were gifted with the ability of extreme synthesis, we could claim that we are in front of a sentimental education that leads us nowhere, a Bildungroman without the forming process and, obviously, without the novel: an album with no memories. We are, however condemned to dwell, or even better, to postpone. As a result of the impossibility to succeed in completing the self, words, icons, descriptions, prefaces, notes, supplements and quotes emerge as wrecks left by the sea on the beach of the imagination. They are writing tools that encircle a void and that create, at the same time, an occasion for communication that is concise but not elementary, primary but not primitive, residual but yet amenable to exert a tautological force.
>
> It is as if the author commissioned himself a work that he cannot possibly finish. He outperforms himself. The only way out would be to draw a line. As a matter of fact that is what usually happens. Any author, at some point, does it. But it is not enough. The buyer is not satisfied. The buyer rejects the work and refuses to pay for it. Filling up empty pages in a perhaps aesthetically sophisticated way is not enough. At the end of the rather entangled relation between the author and the buyer, the novel is left abandoned, half completed, unfinished, still looking for a reader to be adopted and saved from sinking into oblivion. The reader is also a creator who might rapidly bury both author and buyer, either for (neurotic) interests or (perverse) pleasure. At the end of author and buyer's funeral, in the spare time, the text returns to life with demands or even pretensions of being considered a text after all.
>
> It is true that even the washing machine instructions are a text from a semiotic perspective, but our text is a different case. Unfortunately enough, the object in question is born as an orphan and is in poor health; it has suffered from plain, yet not serious, neurological damages which has hampered its navigation skills. This handicap is declared with honesty in the title '*Ataxic Almanac*': a poor comfort for the average public who understandably expects healthy and possibly attractive texts.

The brain is important but it doesn't represent everything: the eye should also be pleased. The Almanac therefore, has to find the right reader, even if the reader will be 'customized' with a sort of handcraft sociology of cultural consumption, a 'nouvelle cuisine' mixing individual skills.

Looking at it fairly, the average Almanac reader has many shortcomings. She is lazy and messy. She does not have the elementary knowledge that only a classic education can provide. She is a passionate television consumer and a compulsive reader of any kind of text. She is gifted with a distinct, yet useless, tendency to classify everything. Her sexual life is poor but not dull. Finally, she has an ironic, almost gruesome, sense of humour. You could not define her as a misfit since she has a social role in life and, furthermore, worries about the future of her community. These latter elements might work in her favour, if only it were not all in vain. On the contrary, the friendly reader, who has the quality of being able to isolate from the author and the buyer, should not be equipped with a few mandatory reading tools: an Ariadne's thread to help her walking out alive from an ataxic labyrinth, a compass to navigate in a sea of details, a medicine to bear the weight of usefulness.

Unfortunately, as it happens, we are not up to the task. Therefore, we will suggest one last example. Imagine yourself in your home (or in your parents' house, if they are still alive). As in every house, there is at least one messy drawer, some houses have more than one (they have, in fact, messy trunks, rooms and cellars). Inside the messy receptacle, you'll find receipts, unpaid bills, Polaroid pictures, business cards, cultural association memberships, unknown phone numbers, letters, postcards, Christmas wishes, bus tickets, notes taken during train trips, novel plot lines written on three-star hotel headed papers, old poems written on anything, birthday candles, French francs, old Italian liras, buttons, safety pins, half-used organizers, diet pills, cheque counterfoils and holy pictures.

The Almanac comes into existence by throwing the drawer content onto the kitchen table. Then you organize such a scattered collection of memorabilia in a vague chronologic order. But there is an essential difference between the messy drawer and the Almanac. The first one is alleged to have an owner. The Almanac has none and yet why should it have? By its very existence the Almanac suggests to us a possible owner. The material the Almanac is made of continuously reminds us of a putative *self*. The materials inside the Almanac crave for a self, but I guarantee you that the only thing to be sure of is that they are lying.

In 1995, Leo Sarastro wrote about nothingness with unabashed severity:

the echo of muffled cries coming from these works ought not to be underestimated. The cause of the sorrow is unknown and the author doesn't help to find it out. He walks on a suspended wire. Wherever he falls, his future is doomed. There is no hope of getting to the other side: it's too far. Time plays a crucial role. Galletta, however, doesn't appear to worry about it. He waits for the fatal false step with commuters' uneasy confidence. He observes the faces of his travel companions and is moved by the fun he gets from it.[1]

---

1   Translated by Chiara Moroni.

References

Adorno, T.W. (1954), *Minima Moralia*, Turin, Einaudi.

Barthes, R. (2003), *La Préparation du Roman I et II: Notes de Cours et de Séminaires au Collège de France 1978–1979 et 1979–1980*, Paris, Éditions du Seuil — IMEC.

Galletta, G. (alias Leo Sarastro) (1995), *Delirium Stabile*, Genoa, Galleria Il Capovolto.

Galletta, G. (alias Leo Sarastro) (2003), *Almanacco di un Altro Anno*, Genoa, Antilibro-posteditore.

Galletta, G. (1981), 'Autopresentazione', first published in *Ghenligures*, Lecce-Genoa. Officially published in G. Galletta (2010), *Il Museo del Caos*, Genoa, Il Canneto editore.

Galletta, G. (2005), 'Lo schema dell'ispirazione' (Inspiration Plan), *Collage Digitale*, cm 21x29,7.

James, W. (1909), *A Pluralistic Universe*, Cambridge (MA), Harvard University Press.

Kraus, K. (1987), *Detti e Contraddetti*, Milan, Bompiani.

Rilke, R.M. (1974), *I Quaderni di Malte Laurids Brigge*, Milan, Garzanti.

Risso, E. (2001), a cura di, 'Intervista a Edoardo Sanguineti' in *Atlante del Novecento Italiano*, Lecce, Manni editore.

Index

3D 96, 206
3D technology 90
3D world 90

Active externalism 5, 37-50, 58
Ada 205
Adams, F. 18, 20, 42, 46
Adi, P. G. 9, 211, 215-16
Adorno, T. W. 234, 236
Aesthetic experience 1-9, 15-18, 51-5, 59, 89, 107, 111, 114, 123-137, 141, 150, 173-5, 178, 180, 186, 189, 211, 214-16, 224-5, 228
Aesthetics 1-10, 15, 31, 37-8, 47, 49-53, 89-90, 100, 108, 123-8, 130-3, 136, 141-2, 146, 206, 211, 214-15, 224, 230
Affordances 28, 73, 132, 164, 166, 215, 225
Agency 56-9, 63, 76, 110, 125-8, 132-3, 221-5
AI / Artificial Intelligence 6, 87, 126, 209, 216, 220-1
Aizawa, K. 18, 20, 42, 46
AL 126
Alexitimia 211, 215-17, 223
Algorithm 190, 203
Almanac 233, 237-8
Alter-ego 10
Amusia 76-8, 80, 82
Analytical aesthetics 37-8, 47, 50-1, 127
Andre, C. 53
Anthropology 121-6, 131, 236
Apollo 213
Archaeology 7, 123-6, 131
Aristotelian 99-100, 156, 185
Aristoxenus of Tarentum 149
Arslan, B. 198
Art 3, 15, 17, 31, 37-61, 69, 87-103, 107-122, 123-6, 135, 141, 159, 166-9, 174, 206, 211-17, 223-4, 226, 228, 230, 234-6
Art consumption 47-50, 112
Artaud, A. 211-12, 227

Artists 7-10, 48, 53-8, 94, 97, 100, 107-17, 120, 168, 173-4, 190-1, 211, 214, 216-17, 220, 223-4, 226, 230, 233
Artworks 5, 7, 9-10, 41, 48, 51-9, 90-4, 98, 100, 108, 110-17, 120, 124, 126-7, 134, 211, 214-7, 221, 223-4, 226, 228
Artworld 53-4, 112, 123-4
Ascott, R. 113-14
Attention/Attentional 8, 44, 51-8, 70-8, 81-2, 117, 128, 132, 142, 145, 148, 159, 162, 197
Auditory experience 6, 63-73, 78-82, 203

Bach, J.S. 69
Barthes, R. 151, 235
Bartok, B. 190
Bateson, G. 26, 28, 120
Beardsley, M. 5, 52, 55, 127, 131
Bell, C. 15, 52, 124
Benveniste, E. 149
Bergson, H. 162
Bernstein, L. 8, 175-7, 181-3, 186
Binding / Boundaries 22, 123, 129-130, 133
Binkley, T. 53-4
Biosignal 197-8, 203
Block, N. 21
Body 2-4, 18-31, 37-8, 44-5, 49, 64, 70, 73-5, 77, 80-1, 88, 90, 96, 117, 121, 128-33, 141, 151, 159-60, 164-9, 180, 194, 206, 211-19, 223, 225, 227-8, 230, 234, 236
Bourriaud, N. 113
Brahms 183
Brains 3, 7, 10, 16-29, 38-41, 44-5, 49-50, 53, 64, 87-90, 107-8, 114, 120-1, 123, 127-9, 133-5, 158, 167, 170, 174, 177-8, 180, 183, 197, 204, 206, 211, 215-17, 219, 221-8, 230, 237
Braintenberg 195
Brentano, F. 93
Brooks, R. 26
Brownian Noise 190, 203

Buddhism 118-9
Bullough, E. 51
Burge, T. 27, 30, 39
Burnett, R. 113

Cage, J. 195
Caravaggio 49-50
Carroll, N. 53-4, 124, 167
Causal 6, 22-6, 37-9, 42, 64, 108, 129, 132, 163
Causality 227-8
Causation 33, 125
Cave art 48, 155-70
Cave writing 155-70
Chalmers, D. 4, 22-3, 37, 40-3, 129
Chambers, D. 41, 47
Chemero, A. 28, 43, 47
Chomsky, Noam 8, 87, 175-6, 181-6
Churchland, P. 22, 26, 52, 184-6
Cinema 234
Cithara 213
Clapton, E. 182
Clark, A. 2, 4, 24, 27, 30, 39-43, 215, 223
Cognition 4, 7, 9, 19, 21, 23-4, 27-30, 41-2, 53, 88, 90, 117, 120, 128, 177, 184, 189, 194-5, 198, 200, 221, 228-9
Cognitive boundary 42
Cognitive Image 88, 113
Cognitive mind 4-5, 23, 27
Cognitive processes 2, 5, 27-8, 38-46, 55, 88, 92, 111, 184, 199, 202
Cognitive systems 32, 60, 118, 194
Cognitive world 8, 169
Collingwood, R. G. 109, 110, 114
Compositional processes 9, 190
Computer based music composition 9, 189-91, 195, 197, 199, 201, 203-4
Computer based music system 9, 189, 191-8, 200-1, 205
Connectionism 187
Consciousness 16, 21, 24, 26-9, 64, 69, 88, 109, 111-14, 117, 119-21, 123, 156, 158, 161-2, 164, 167-9, 179, 211-12, 221-8
Contemporary art 48, 51-9, 191, 217
Contemporary literature 8
Content externalism 25
Contextualism 37, 49-7, 124, 143, 145, 161, 200
Creative flux 225
Creative processes 127, 136, 228
Creative skin 212-3
Creativity 9, 111, 120, 136, 164, 167, 189, 196, 236
Crick, F. 21, 90
Csikszentmihaly, M. 196
Cubism 52
Cupchik 76
Currie, G. 1, 56, 58-9
Curvasom 200-1
Cutaneous secretion 9, 211-12, 230

Cyborg 59, 223

Danto, A. 5, 15, 51, 53-9
Davies, D. 56-7
Debussy 69, 190
Deep listening 64, 67, 70-8, 80, 82
Deep structure 74, 176-7, 181
Deleuze, G. 131, 212, 228-9
Dennett, D.C. 17, 27, 30
Denora, Tia 78
Deronda Daniel 158-61
Derrida, J. 166
Descartes/Cartesian 1, 5, 29, 41, 129, 221-22, 229
Developmental robotics 226
Dewey, J. 26, 29, 124, 179
Diary 235
Diatonic harmony 182
Diatonic scale 8, 175-6
Dickie, G. 1-2, 15, 53, 112, 131
Digital literature 8, 155, 165-6, 169
Digital writing 166-7
Directionality 72-3
Distant touch 217
Dopamine system 204
Dostoevsky, F. 162
Dramatis personae 157
Dretske, F. 27, 30
Du Champ, M. 48, 53-4, 91, 111-12, 224
Dutton, D. 56
Dynamical systems theory 221

Ecological view of perception 19, 26, 93
Ego 10, 94, 177-8, 180, 229, 233
Egocentric 65-6, 70-1, 76, 96-9
Electrocardiogram 226
Electrode 203, 218-23
Electroencephalogram 197, 203
Eliot, G. 158-61, 169
Embedded mind 32-4
Embodied mind 30, 32
Embodiment 3, 7-9, 19-20, 24, 28, 128, 131, 163, 189, 194, 201, 211, 213, 215, 230
Emin, T. 53
Emotional 26, 71, 77-81, 92, 109-10, 144, 156, 158, 160-1, 164, 167, 174, 187, 189, 196-8, 201-5, 215
Emotions 77-8, 117, 158-60, 174, 186, 189, 197-8, 200, 204-5, 214-15
Empathy 160, 214-5, 224, 228
Enactivism 5-6, 21, 28-30
Enchantment 124
English literature 8, 159
Eno, B. 206
Environment/Environmental 2-6, 18-20, 23-9, 39-40, 42, 44-5, 48, 56, 64-7, 73, 78-9, 82, 92-4, 108, 114, 117, 120, 128, 135-6, 155-6, 159-60, 162, 165-6, 168-9, 178-9, 189,

194-5, 199-200, 203-6, 211-12, 217, 221, 223, 225, 227-9, 236
Euripides 144
Evil genius 222
Evolutionary robotics 226
Experimental aesthetics 90, 100
Experiments in Musical Intelligence 191
Explanatory externalism 5, 37, 45-51, 109
Expressionism 52
Extended mind 4, 24, 27, 30, 39, 41, 88, 111, 120, 215
Extended self 173-187
Extended space 6, 68, 87-106
Extensionism 7, 107-122
Externalism 2-10, 15, 17, 19, 21-31, 37-51, 58-9, 87-92, 131, 155-9, 161, 168, 189, 206, 211, 215, 222

Fechner, T. 90
Fibonacci 190
Flaubert, G, 162
Fodor, J.A. 22, 46, 87, 194
Fowles, J. 165
Fraunhofer, S. 197
Freud, S. 110, 159, 177-8, 180, 185-7
Fry, R. 15, 52

Galileo 89
Gallagher, S. 16, 26, 28, 87
Galsworthy, J. 161
Gell, A. 110-111, 117, 124-6, 131
Gestalt 88, 93, 214
Gibson, J. J. 26, 28, 43, 87-8, 95, 161, 168, 215, 225
Goethe, J. W. Von 147
Gombrich, E. 109, 110, 149
Gravity 25, 55, 82, 174
Greek mythology 213
Greenberg, C. 52, 55-7
Guattari, F. 131, 212

H₂O 39
Hamilton, A. 65, 217
Hard problem 22
Haslbeck, F. 78-80
Hearing 6, 56, 64-5, 67, 69, 71-2, 78-9, 179, 182-3, 213
Heart-rate 197, 203
Heidegger, M. 130, 179
Hekkert, P. 41
Hering 93
Hildebrand, A. 97-8, 100
Hill, C. 113
Hobby-horse 156-7, 163, 169
Holt, E. 25-6
Honderich, T. 29-30, 158
Horkheimer 234
Hume, D. 108, 175, 177-8, 180
Hurley, S. 21, 23, 28, 30, 42-7, 63

Hutcheson, F. 54
Hutchins, E. 120, 203-4
Huxley, T.H. 159

IanniX 17
Identity theory 174
Idiolect 147
Ihde, Don 68-9, 71-2, 78
Imitatio naturae 156
Impressionism 52
Information 9, 26, 66-7, 82, 88-9, 92-3, 95-9, 109, 128, 135-6, 158, 160-1, 166, 169, 174, 178, 183, 189, 191, 194, 196-8, 202, 205-6, 214, 219, 225-6, 229, 236
Ingold, T. 124, 129-31
Institutional(ism) 51, 53, 92, 112
Integration(ism) 19, 41, 43, 149
Intention(al) 18, 23-4, 69, 72, 111, 124-29, 132, 134, 144, 150, 198, 214, 223
Intentionality 18, 27, 64, 125, 132
Interactive music system 9, 189, 191-201, 206
Interactive setup 9
Internal self 233
Internal structure 174-5
Internalism 5-6, 18-21, 24, 29-31, 37-9, 42-3, 46, 51-5, 87-93, 134-5, 158-9, 169
Interpretation 20, 24, 41-2, 51-6, 88, 112, 124, 158-60, 163, 166, 182, 194-5, 198, 220
Intuition 1, 4, 30, 146, 149, 164
Iseminger, G. 15
IXI 197

Jackson, F. 27
Jakobson, R. 148-9
James, H. 159, 162
James, W. 179, 233

Kandinsky 94
Kant 15, 93, 108, 174-5, 177-8
Kehlmann, D. 146-7
Kierkegaard, S. 55, 68-9
Kim, J. 17, 21, 33
Kindy, J. 49
Klages, L. 148-51
Klee 94
Koch, C. 15-19, 21, 24, 28, 90
Koffka 94, 99
Kohler, W. 89-90
Kraus, K. 236
Krauss, R. 229

Lacan 166
Lakoff, G. 175, 181-2
Language 1, 7-9, 41, 46, 68, 88, 125, 141-53, 155, 159, 165, 173, 175-7, 181, 183, 191, 211-17, 222, 236
Language of emotions 189
Law, J. 132
Lawrence, D. H. 161

Leonardo da Vinci 94
Lewis, C. S. 173, 186
Lewis, M. 79
Lewis, P. 162
LeWitt, S. 57-8
Linguistics 7, 27, 39, 41, 88, 91-2, 141-9, 152, 165-7, 170, 175-6, 181, 183-4, 189
Lippard, L. 57
Literature 8, 17, 24, 63, 68, 78, 147, 155-70, 173, 197-8, 205, 234-5
Livingstone, M. 50, 90, 107
Location question 5, 17-19
Locational space 66-7, 70-1, 73-6, 82, 112
Locke, J. 156
Logos 213
Long, R. 57-8
Louis, M. 52, 55-6
Lycan, W. 23, 27, 30

Mach, E. 180
Machover, T. 192, 197
Margolis, J. 57
Markov Chains 190, 203
Marquesan art 110-111
Marr, D. 26, 88
Marsyas 213
Martindale, C. 107
Martinet, A. 145
Marussich, Y. 217
Massys, Q. 109
Material engagement 7, 123-139
Maturana, H. R. 87-8, 225
MEART 9, 211, 217-226
Melzack, R. 20
Memory 40-3, 56, 100, 120, 136, 149, 163, 194, 200, 202-3, 221, 233, 236
Menary, R. 24, 40-2
Mental processes 5, 37-40, 45, 159
Merlau-Ponty, M. 28, 69-70, 87, 227-8
Methodological philistinism 124-5
Metzger, W. 94, 97
Michotte, 96, 100
Microconsciousness 24, 228
Microtones 183
MIDI 198, 200-202, 205
Mimesis 156, 168
Minimally sufficient neural substrate 21, 134
Modern literature 161, 165-6, 169
Modernism 52-3, 57, 59
Modernist painting 52, 55-7
Modernist writing 161, 170
Modulation 74, 81, 182, 191, 202, 206
Mol, A. 132
Mondrian 50
Monty Python 186
Morgan, R. 69, 72
Morpheme 8, 176
Mrs Dalloway/ The Hours 162-4
Muller 20

Muscles 4, 24-5, 184
Music 6, 63-85, 107, 142, 149, 168, 173-87, 189-211
Music composition 9, 113, 189-210
Musical machine 189, 191
Musical space 6, 67, 69-73, 77
Musicking 64

Narcissism 234
Nauman, B. 57-8
Neural correlates of consciousness 90, 134-5
Neural networks 9, 20, 180, 184, 186, 190, 194, 221, 227
Neural processes 2, 9, 18, 22, 107, 158, 211
Neuroaesthetics 7, 49, 53, 90, 107-8, 114, 121, 123, 128, 134, 214
Neuromatrix 20
Neurons 4, 16, 24, 90, 134, 185, 214, 217-221, 226-7
Neuroscience 1-2, 4-6, 10, 15-16, 19-22, 28-30, 87, 90-1, 107, 128, 134, 189, 204, 224, 226
Nietzsche, F. 147-8
Noe, A. 21, 28, 30, 45, 63, 67, 120, 135, 160, 166-7, 226
Noland, K. 52
Nonken, M. 68
Novel 146, 148, 155-170, 175, 181, 235-8

O'Callaghan, C. 66, 68
O'Shaughnessy, B. 64-66
Oliveras, P. 67
OMax 191
Optical flow 26
Orlando 164
Otto 40

Pachet's Continuator 191, 196
Pain 20, 173
Panpsychism 34
Parity principle 5, 39-45
Peirce, C. S. 8, 142-3, 149-50
Perception 6-8, 15, 19, 22, 26, 28, 37, 43-5, 48-9, 54-9, 63-8, 72-7, 82-3, 87-103, 128, 130, 132-4, 136, 141-2, 144, 146, 150, 158-63, 166-7, 178, 184, 189, 191, 194-5, 198-202, 206, 213, 222-8
Perceptual consciousness 29, 64, 167
Perec, G. 148
Performer 74, 125, 157, 166-8, 183, 192, 203-6
Personal identity 234
Pettit, P. 27
Phenomenal experience 4, 6-8, 16-7, 20-5, 29-30, 51, 90, 147
Phenomenal externalism 2, 27, 29-30
Phenomenology 5-6, 16, 29, 64-71, 76, 82, 87, 90, 127, 132, 156, 219, 222-3
Philips-Silver, J. 74, 80
Phoneme 8, 145, 176

Photography 48, 185, 218
Physicalism 2, 6, 17, 19, 21-22, 29-30
Picasso, P. 114-9
Pictorial representation 6, 55-6, 91, 94-9, 214, 217
Pictures 88, 93, 95-6, 98, 109, 229, 238
Pirandello, L. 235
Plato 145
Poetry 8, 146-7, 149-52, 164, 176, 238
Pollock, J. 132
Postmodernist literature 8, 165-6, 169-70
Potter, S. 217, 210-11, 225
Pottery 7, 123, 127, 129, 132-6
Prinz, J. 21, 24
Processes 2-6, 9, 16-17, 19, 22, 26-30, 37-45, 68, 71, 78, 88, 91-2, 99, 107, 109, 111, 116, 128, 130, 132-4, 142, 150-1, 159, 162, 165-9, 185, 189-90, 194-5, 199, 202-7, 211, 213, 221
Proust, M. 69, 235
Psychology 1-2, 46, 53, 64, 10, 178, 202, 204, 234
Putnam, H. 27, 30, 39, 64, 222

Qualia 21, 27, 31-2, 164

Radical externalism 29-32, 102, 122
Ramachandran, V. S. 15, 52, 90, 107, 214
Rat neurons 217-8
Reflexive monism 27-8
Reich, S. 195
Reisberg, D. 41, 47
Relational aesthetics 113
Relational art 112-3
Relaxation 197
Representation 3, 6, 9, 15, 24-9, 43-4, 52, 67, 70, 72, 87-90, 94-5, 98, 123-4, 127, 130-1, 134-6, 143-6, 155, 159, 161, 168, 178, 189, 192, 194, 198-9, 202, 205, 211, 217, 219, 222-9
Revonsuo, A. 21-2
Rhinehart, D. 182
Rhythm 7-8, 67-70, 73-82, 141-53, 168, 198, 202-5, 226
Riley, T. 195-6
Rilke, R. M. 146, 235
Roboser 200-1, 205-6
Robot/Robotics 126, 189, 193, 195, 200-1, 205-6, 216-221, 226
Robotic artist 9, 211
Rockwell, T. 8, 22, 29, 228
Rokeby, S. 197
Rolland, R. 178, 180
Rothko paintings 49
Rousseau, J.-J. 159
Rowe, R. 191-3, 203, 205

Sanguineti, E. 236
Santana, C. 182
Sarastro, L. 233, 237-8
Schonberg, A. 175
Schopenhauer, A. 68
'Screen' 167
Seeing 7-8, 44-5, 57, 59, 94-5, 97-100, 111, 115, 117, 127, 129, 133, 176-7, 206
Self 9-10, 20, 43-4, 56, 65, 87, 113-14, 149, 156, 159, 162-4, 173-87, 233-9
Semantic composition 175
Semantic content 2, 22-3, 27, 30, 148, 176
Semantic externalism 2-3, 26, 30, 222
Semantics 18, 25, 28, 157, 205
Sense of learning 149, 152
Sense of order 149, 150
Sense of reaction 78, 142-3, 152
Sense of reality 95, 142, 155, 157, 163, 168
Sensory qualities 89
Seth, A. 21
Shakespeare, W. 150, 186
Shannon, C. E. 88, 214, 226
Shearman, J. 109
Sheldrake, R. 120
Shelley, J. 54
Sibley, F. 54
Silent barrage 219-224
Situated aesthetics 3-7, 126, 130-3, 141-2, 146
Situated body 133
Situated robotics 189
Situatedness 7-8, 24, 72, 133, 143-6, 150, 152, 189, 195
Skin 1-11, 24, 27, 29, 39, 111, 115, 117, 127, 129, 135, 141-2, 159-61, 164, 169, 211-17, 227, 230
Skov, M. 107
Sloboda, J. 64
Smell 6, 89, 93, 117, 213
Smuse 190, 201-6
Solso, R, 90, 107
Sonic patterns 67, 74-5
Speech 8, 39, 56-7, 76, 80, 142-52, 157, 176, 178, 183
Spread mind 29
Sruti 183-4
Sterne, L. 155-8, 161, 165, 169
Stevens, W. 1, 233
Stigmergy 203
Stolnitz, J. 51
Stravinsky 190
Subjective / Subjectivity 6, 8, 51-2, 68, 78, 91, 93, 107-8, 111, 119-21, 124, 126, 143, 149-56, 158-9, 166, 174-8, 186, 196, 225
Subjective experience 52, 108, 121, 145
Supervenience 4, 21-22, 25
Surface structure 176
Surroundability 72-3
Sutton, J. 40-1, 204
Synthetic aesthetics 207
Synthetic interactive music 189, 201
Synthetic method 199

Taste 6, 10, 89, 107-8, 116, 124-6, 141-2, 213
Temporality 68-9
Tetris 40
Textural qualities 67, 80
Thompson, E. 21, 30
Tillmann, B. 76
Tiravanija, R. 113
Titchener, E. 178
Tonneau, F. 21, 26, 30, 120
Touch 6, 44, 63, 68, 83, 93-8, 116, 133-6, 164, 213-17
Trainor, L. J. 74, 79-80
Tristram Shandy 155-9, 163-4
Turner, J. M. W. 95
Twin earth 23, 39

Ulysses 234

Value of art 51-4
Van Leeuwen, C. 41
Vanouse, P. 223-4
Varela, F. J. 30, 87-8, 91, 225-8
Vastfjall, D. 71, 77
Vehicle externalism 2, 23-7, 30, 32, 40, 73, 75, 136, 195

Velmans, M. 27, 120
Verstijnen, I.M. 41
Vico, G. 199
Vision 6, 44, 52-3, 63, 65, 83, 88-9, 91, 93-8, 100, 107, 134-5, 226-7
Visual field 92
Visual space 91, 95
Von Helmholtz, H. 89, 93-4, 98
Von Humboldt, A. 146-7
Von Humboldt, W. 145

Waisvisz, M. 197
Warhol, A. 54, 234
Water 2, 25, 39, 107, 186, 216
Weaver, W. 88
Whitehead, A. N. 17, 26, 119, 121
Wightman, F.L. 66
Winkler, T. 191-3
Wittgenstein 1, 181
Woolf, V. 161-4, 169
World-making 69

Zahorik, P. 66
Zeki, S. 2, 15, 24, 53, 90, 107, 135, 228
Zeus 213